PANDEMIC

THEOLOGY

SAINT SEBASTIAN INTERCEDING FOR THE PLAGUE STRICKEN

Josse Lieferinxe, 1497-1499

PANDEMIC

Jamela Camat
Jonathan Wiebe
Paul Dansereau

Lucas Tombrowski
Austin Mardon
Catherine Mardon

THEOLOGY

A Golden Meteorite Press Book. Printed in Canada.

© Copyright 2020, Jamela Camat, Jonathan Wiebe, Paul Dansereau, Lucas Tombrowski, Austin Mardon, & Catherine Mardon.

Golden Meteorite Press, Edmonton. All rights reserved for Pandemic Theology ©. No part of this publication may be reproduced, stored in any retrieval system, or transmitted in any form or by any means, electronic, mechanical, photo-copying, recording, microfilm reproduction and copying, or, otherwise, without the prior express written permission of Golden Meteorite Press:

First Printing: 2020

Design and Format by August Schaffler

Telephone: 587-783-0059
Email: aamardon@yahoo.ca
Website: goldenmeteoritepress.com

Additional copies can be ordered from: Suite 103 11919-82 Street NW Edmonton AB T5B 2W4 CANADA

ISBN 978-1-77369-145-9

Like the Good Samaritan, may we not be ashamed of touching the wounds of those who suffer, but try to heal them with concrete acts of love.

— POPE FRANCIS —

CONTENTS

Chapter I.   *Introduction*
II.   *Ancient and Medieval Christian Theology*
III.   *Ancient and Medieval View of Medicine*
IV.   *The Christian Vocation to Love*
V.   *The Black Death*
VI.   *The Black Death and Christianity*
VII.   *Spanish Flu*
VIII.   *Spanish Flu and Christianity*
IX.   *Famines in Recent History*
X.   *Famines and Christianity*
XI.   *Modern Christian Theology*
XII.   *HIV/AIDS*
XIII.   *HIV/AIDS and Christianity*
XIV.   *COVID-19*
XV.   *COVID-19 and Christianity*
XVI.   *Conclusion*

I.

INTRODUCTION

If there's anything that the recent COVID-19 pandemic has succeeded in, it's reminding us of how terrifying pandemics really are. Imagine: an unseeable agent spreading from person to person, seeming to ignore all borders, both physical and societal; causing debilitating symptoms and lifelong disabilities when it doesn't result in outright death. Today, we have the benefit of being able to scientifically identify and even visualize the viruses or bacteria responsible for epidemics, and many of those in developed countries live with the insurance that comes with robust health care systems and the latest in medical and pharmaceutical technologies. Now, imagine living in a time where people didn't have those kinds of assurances, and the vast hopelessness that would be inherent in living through a disaster that terrifying. This book will discuss the histories of some of the most deadly and influential pandemics in human history, along with contemporary Christian responses and modern theological perspectives on these pandemics. Besides the history of pandemics, we will provide general overviews on topics such as ancient and medieval medicine and the history of Christian theology from ancient to modern times, in order to provide a more robust picture of the history of pandemics and Christianity. Along with pandemics such as the Black Death, the Spanish Flu, HIV/AIDS, and COVID-19, we will reflect on the religious and historical reactions to notable famines within the last several centuries, highlighting the 'man-made' aspect of global disasters.

A purely secular summary of historical pandemics would be completely adequate for providing raw historical facts, analyses, and consequences of each pandemic in question. That being said, it can be very useful to examine such important events from a theological perspective, especially for a person of faith. This book may focus on Christian theology and perspectives on pandemics that had notable effects on Christianity or responses from Christians, but the importance of historical spiritual reflection is important for any religion. Through examination of history, we can gain a greater understanding as to why Christian faith is what it is today, and by looking at the positive and negative reactions people had at the time we can often see examples of what we should or shouldn't do, not just during a pandemic, but at any time in our lives when people are facing great suffering. Our goal is not to attempt to answer age-old theological questions about the role of evil, death, and hardship in God's world, but rather to identify the "ideal" Christian response to widespread suffering. To this end, we will discuss the Christian vocation to love, an essential component of Christian life which provides an avenue for providing unconditional love for other people.

II.
ANCIENT AND MEDIEVAL CHRISTIAN THEOLOGY

>
> *THE CHRISTIAN FAITH PRESUPPOSES CERTAIN TRUTHS OF REASON AND OF HISTORY. WITHOUT THESE IT CANNOT MAKE SENSE AND ITS THEOLOGY CANNOT BECOME AIRBORNE. FIRST OF ALL, THE CHRISTIAN FAITH PRESUPPOSES THAT THERE IS A GOD. IT IS NO USE TELLING PEOPLE THAT JESUS IS THE SON OF GOD IF THEY DO NOT BELIEVE IN GOD IN THE FIRST PLACE.*[1]

[1] Aidan Nichols, The Shape of Catholic Theology: An Introduction to Its Sources, Principles, and History (Collegeville, MN: The Liturgical Press, 1991), 55.

It is fairly safe to say all religions are based upon certain presuppositions. With regards to Christianty, as the above quote illustrates, there are a number of foundational presuppositions. In this chapter we will take a brief look at how certain aspects of Christian theology evolved, stemming from certain presuppositions, starting with the years immediately following the death and resurrection of Jesus until the death of the last of the apostles; when Christianity was just beginning to take root and Christian theology had yet to distinguish itself from Judaism which preceded it, commonly referred to as the Apostolic period. Next, we'll see some of the developments made during the Patristic period, which is considered to have lasted from the end of the Apostolic period until roughly 600 A.D.. Not only did Christianity first become legal at this time, but having inherited much of classical Greco-Roman culture, it also preserved much of it following the collapse of the Roman Empire and set the stage for the subsequent rise of Christendom during the Medieval period. Theologically speaking, the contributions of the Patristic Fathers can hardly be overstated. Finally, we'll take a look at Scholasticism, which peaked around the 13th century, and the monumental influence it had on all subsequent Christian theology up until today. Even though it is impossible to do justice to all the theological developments which took place during these years, we will attempt to sketch a rough picture of the distinctly Christian worldview which emerged, setting the stage for understanding how Christianity and the Church have responded to pandemics up and down the centuries. In order to understand Christianity, one of the first things to establish is its inherent connection with Judaism.

> "BOTH JUDAISM AND CHRISTIANITY HOLD THAT THE ONE TRUE GOD, SOURCE AND GROUND OF THE WORLD, HAS DISCLOSED HIMSELF THROUGH THESE EVENTS OF RELIGIOUS HISTORY—FROM ABRAHAM TO THE CLOSE OF THE OLD TESTAMENT PERIOD IN JUDAISM, FROM ABRAHAM TO THE APOSTLES OF JESUS IN CHRISTIANITY. THE CREATOR LORD HAD ENTERED A COVENANT RELATIONSHIP, A RELATIONSHIP OF FRIENDSHIP AND TRUST WITH A PEOPLE: ISRAEL, THE CHURCH."[2]

[2] Ibid., 79.

Even though the Christian worldview shares many of the same basic tenants with Judaism, it does have a number of distinctive features. For example, Christianity believes that the Hebrew Scriptures do not contain the fullness of divine revelation, but that they are essentially incomplete, containing an unfulfilled promise, if you will, comparable to a story without a climax. Judaism at the time of Jesus was one of intense expectation, awaiting the fulfillment of the Old Testament promises, which is why all four Gospel accounts mention John the Baptist as a precursor to Jesus, "the one of whom the prophet Isaiah spoke when he said, 'The voice of one crying out in the wilderness: Prepare the way of the Lord, make his paths straight.'" (Mt 3:3); John is a bridge between the Old and New Testaments, announcing that Jesus is the fulfillment of everything they were waiting for. What we find in the three Synoptic Gospels is, in many ways, an account of how, exactly, Jesus fulfills the Old Testament prophecies, even though not necessarily in the way they were expecting. As he commences his public ministry with the words: "The time is fulfilled, and the kingdom of God has come near; repent, and believe in the good news." (Mk 1:15); Jesus is proclaiming the lordship of God, telling us: "God is acting now—this is the hour when God is showing himself in history as its Lord, as the living God, in a way that goes beyond anything seen before." [3] Meanwhile, the Gospel of John on the other hand presents Jesus as the Divine Logos, the "Word made flesh" (Jn 1:14), sent by God to make it possible for humanity to share in the very life of God: "For God so loved the world that he gave his only Son, so that everyone who believes in him may not perish but may have eternal life." (Jn 3:16, NRSV) In one sense, all four Gospels do tell the same story and convey the same message, however, at the same time: "To the careful reader, each Gospel does put a different slant on Christ, and so on its entire presentation of the Christian mystery." [4] A good illustration of this is that the phrase "Kingdom of God" is all but absent from the Gospel of John, however, the concept is still present, having been more-or-less replaced with the proclamation of "eternal life".[5] That being said, if the reader understands what is meant by each of these phrases, it is possible to see the common themes being explored by each of the Evangelists. Another example being the call to conversion—metanoia in the Greek, to "go beyond the mind that you have" [6] —which is a common theme found in all four Gospels. What the four Evangelists and other New Testament authors understood is that God was indeed acting "in a way that goes beyond anything seen before", therefore, it requires a new mindset and a new lens with which to view the world.

[3] Joseph Ratzinger (Benedict XVI), Jesus of Nazareth: From the Baptism in the Jordan to the Transfiguration (New York, NY: Image Publishing, 2007), 56.

[4] Nichols, 268.

[5] See the context of John 3:16, Jesus' conversation with Nicodemus in John 3:1-21, for an example of this.

[6] Bishop Robert E. Barron, "Why Having a Heart of Gold is not what Christianity is About," Word on Fire, January 28, 2015, https://www.wordonfire.org/resources/article/why-having-a-heart-of-gold-is-not-what-christianity-is-about/4665/

"AS WE MOVE INTO THE PATRISTIC AGE, WE ARE MOVING INTO A SECOND ORDER OF REFLECTION. THEOLOGY IS

> REFLECTING ON THAT PRIMARY OR IMMEDIATE REFLECTION WHICH IS THE NEW TESTAMENT. IT IS LOOKING BACK, TAKING STOCK AND ASKING ITSELF WHAT SORT OF A WORLD THIS MUST BE IF ITS CENTRAL REALITIES ARE GOD, CHRIST, THE SPIRIT, THE CHURCH, AS THE FIRST FRUITS OF THE KINGDOM. REVELATION IS BECOMING DOCTRINE."[7]

Not only that, but: "When an essentially Semitic gospel encountered the Hellenistic world, it had to express itself in categories that could be understood by those who belonged to that world."[8] Therefore, even as it was wrestling with a number of important issues, Christian theology also found itself adapting its terminology and otherwise making use of Greek philosophy, primarily in order to express itself in terms that would be familiar to the surrounding culture. Arguably the biggest question theologians were wrestling with at the time had to do with trying to understand how Jesus could be both divine and human, as the Gospels and New Testament clearly presented him, and yet still hold the belief that God is one; the strict monotheism inherited from Judaism. What developed over time were a number of approaches, some emphasizing the humanity of Jesus, others his divinity. In the theological school at Alexandria, for example, the focus was primarily on the concept of the Logos, a term found in Greek philosophy meaning the divine Word or reason, used in John's Gospel to refer to Jesus himself: "In the beginning was the Word, and the Word was with God, and the Word was God." (Jn 1:1) Many of the theologians associated with Alexandria had a great deal of reverence for the transcendence of God and therefore emphasized the divinity of Jesus. Meanwhile, in Antioch, the other major theological school of the time, there was an emphasis on "a Logos-anthropos Christology which emphasized the full humanity of Jesus... 'The core of Antiochene Christology lies in a consistent vision of Jesus Christ as a historical figure or person who

[7] Nichols, 273.
[8] Ibid., 275.

bore two distinct natures.'"[9] The benefit of approaching the problem from two different angles meant that, in this case, not only did they help balance each other, but when it came to answering the challenge posed by a number of prominent heresies, Christian theologians were better prepared to defend what was considered to be 'orthodox' faith. For example, the heresy known as "adoptionism" held that Jesus was the "adopted" son of God; meanwhile "modalism" considered Jesus simply to be one of the "modes" by which God manifested himself— both directly contradicting the doctrine of the Trinity. Arianism on the other hand, denied the eternity of Christ, considering him but "the highest creature, the one by whom God created the other creatures".[10] Eventually all this theological debate led to the Councils of Nicea (325 A.D.) and Constantinople (381 A.D.), where "a decisive criterion for understanding the faith and for guiding its expression" was developed,[11] often referred to as the Nicene-Constantinople Creed. In the process of formulating a definitive statement of orthodox belief, in opposition to the Aryans, for example, the Council of Nicea not only affirmed the full divinity of Jesus but stated that he was 'con-substantial' with God the Father, the first 'person' of the Trinity, and therefore, having been begotten "from the substance of the Father", he is "essentially distinguished from the whole created universe";[12] meanwhile refuting both adoptionism and modalism in the process. The debate continued, however, until the Council of Chalcedon (451 A.D.), which settled things down to some degree.

> "IF NICAEA HAD SAFEGUARDED THE DIVINITY OF JESUS, CHALCEDON PLACED EQUAL EMPHASIS ON HIS FULL HUMANITY, JUXTAPOSING FOUR CAREFULLY BALANCED DOUBLE AFFIRMATIONS WHICH ECHOED THE EARLIER COUNCIL: JESUS WAS PERFECT IN DIVINITY, PERFECT IN HUMANITY; TRULY GOD, TRULY MAN; CONSUBSTANTIAL WITH THE FATHER, CONSUBSTANTIAL WITH

---

[9] Thomas P. Rausch, Who is Jesus?: An Introduction to Christology (Collegeville, MN: Liturgical Press, 2003), 153.

[10] Gilles Emery, The Trinity: An Introduction to Catholic Doctrine on the Triune God, trans. Matthew Levering (Washington D.C.: The Catholic University of America Press, 2011), 62.

[12] Ibid., 70.

[13] Rausch, 161.

US; BEGOTTEN BEFORE THE AGES, IN THE LAST DAY THE SAME FOR US...FROM MARY. IN MANY WAYS THE CHALCEDONIAN CONFESSION WAS A SYNTHESIS OF VIEWS."[13]

It is important to understand how much hinged on these issues. If Jesus was not God, then he was either a liar or a lunatic, and we are not saved but still marred by original sin—for only God can forgive sins. If Jesus was not fully human, then we have no chance at participating in the divine life he claimed to offer: "I came that they may have life, and have it abundantly." (Jn 10:10) The doctrine of the Trinity, original sin and salvation, all depend on these fundamental claims, as does the very existence of the Church and everything else that follows.

Turning to look at scholasticism, which could be described as the result of an intense devotion to the patristic fathers encountering Aristotelian metaphysics and logic—as well as the theological high point of the medieval period—we find a synthesis of faith and reason unlike anything ever seen before. What theologians of the scholastic period sought to demonstrate is that "a profound structural similarity exists between nature and grace. Thus the description of what is involved in natural processes in people can provide a model for understanding what is involved in (the) supernatural."[14] This, in turn, might be thought of as "a fundamental option in the understanding of biblical revelation, an option which emerged at a certain point in the theological history of Catholic doctrine and which, when it had emerged, the Catholic tradition would embrace in a decisive way."[15]

By far the most prominent and influential theologian of this period, St. Thomas Aquinas, was known for his wholehearted embrace of Aristotelian logic:

"ST. THOMAS'S THEOLOGY IS POISED BETWEEN, ON THE ONE HAND, A PEDAGOGICAL, EVANGELICAL,

[13] Rausch, 161.
[14] Nichols, 299.
[15] Ibid.

PASTORAL CONCEPT OF WHAT THEOLOGY SHOULD BE AND, ON THE OTHER, A SYSTEMATIC, REFLECTIVE APPROACH. HE INSISTS THAT IN ITSELF THE REVELATION WE HAVE RECEIVED FROM APOSTLES AND PROPHETS IS BOTH ENTIRELY INTELLIGIBLE AND ALSO COMPLETE."[16]

Meanwhile, on the other hand, medieval Augustinians such as St. Bonaventure and others who followed in the theological footsteps of St. Augustine: "characteristically maintained that reason, while competent to deal with earthly and temporal matters, had no strict rights in what was spiritual and eternal."[17] In addition to the Neo-Augustinians, the theological 'movement' known as nominalism also questioned the validity of Aristotelian metaphysics, however not necessarily for the same reasons.

"THE NOMINALIST THEOLOGIANS WERE SO CALLED FROM THEIR BASIC PHILOSOPHICAL OPTION. THEY HELD THAT HUMAN CONCEPTS ARE NOT GENERATED BY A REAL UNION BETWEEN OUR MINDS AND THE NATURE OF THINGS BUT ARE MERELY NAMES, NOMINA, BY WHICH WE LABEL THE WORLD AROUND US FOR OUR CONVENIENCE. NOMINALISM WAS A CURIOUS COMBINATION OF SCEPTICISM AND PROFOUND RELIGIOSITY, AS WE CAN SEE FROM ITS GREATEST EXPONENT, THE ENGLISH FRANCISCAN WILLIAM OF OCKHAM." [18]

What Ockham and the nominalists sought to do was champion a new concept of freedom; the power to choose between contraries independent of all other causes, as freedom of indifference. This was because Ockham's thought was:

"DOMINATED BY THE IDEA OF DIVINE OMNIPOTENCE, WHICH ENABLED HIM TO CARRY HIS IDEA OF FREEDOM TO AN ABSOLUTE DEGREE. FOR HIM, THE DIVINE WILL

---

16 Ibid., 300.
17 Ibid., 302.
18 Ibid., 305.
19 Servais Pinckaers, The Sources of Christian Ethics, trans. Mary Thomas Noble (Washington D.C.: The Catholic University of America Press), 246.

> WAS TOTALLY FREE; IT GOVERNED MORAL LAW ITSELF AND ALL THE LAWS OF CREATION. WHAT GOD WILLED WAS NECESSARILY JUST AND GOOD PRECISELY BECAUSE HE WILLED IT."[19]

The unfortunate result is that obligation became the centre of Ockham's moral theory and subsequently led to a morality of obligation which confused love with obedience to obligations. In one sense, undermining the commandment to "love thy neighbour", because the focus was not on willing the good of the other, but because it was what God had commanded; and in place of the virtues, obedience was held up as the supreme good. Leading, in turn, to what became known as the manualist tradition, which held sway in Catholic moral theology up until the Second Vatican Council.

By the end of the medieval period, Christian theology had progressed a great deal from when the New Testament was being composed during the first-century. At the same time as they sought to synthesize what they had inherited from Judaism with Greek philosophy, the patristic Fathers offered further reflection on the life and teaching of Jesus Christ, and succeeded in bequeathing a rich Christian heritage to the subsequent generations. Those who came after them built upon this foundation, and made further progress in understanding the inherent connection between faith and reason. As we hope to show in subsequent chapters, thanks to these developments, we are not only able to articulate what exactly constitutes an authentic Christian response to the suffering and devastation brought about by pandemics, but also what might be classified as the exact opposite.

III.
ANCIENT AND MEDIEVAL VIEW OF MEDICINE

> IN FACT, THE HISTORY OF MEDICINE IS LARGELY THE STORY OF POISONS MISCONSTRUED AS CURES, PLACEBO EFFECTS, AND NATURAL RECOVERIES FALSELY ASSIGNED TO WHATEVER THE DOCTOR, WISE WOMAN, SURGEON, DRUGGIST OR QUACK HAPPENED TO HAVE DONE.[20]

As is the case with much of what we have in our world today, modern medicine has come a long way from its roots, with the history of medicine going nearly as far back as recorded human history, if not longer. Over the periods that we have recorded, the ancient understanding of medicine and disease grew and changed greatly, beginning with early work from such famous figures as Hippocrates - where the work he and others of his time did was largely speculative and based on external observations of the human body and sickness - being developed by later figures like Galen, and being developed further still throughout early Christianity and into the Renaissance. Despite how different medicine is now compared to its humble origins, and how far our understanding of diseases has come, it truly wasn't that long ago that humanity was still discussing health in terms of Hippocrates' four humors – remarkably enough, Hippocratic theories on medicine and disease have been the driving force behind medicine far longer than not, even lasting into the beginnings of modernity.[21] With this rich lineage in mind, the aim of this chapter is to present an abridged version of the history of medicine, in an attempt to show

[20] "Introduction to Late Medieval Medicine," Michigan State University, https://history.msu.edu/hst425/resources/online-essays/introduction-to-late-medieval-medicine/.

how our understanding of medicine and disease grew and developed over time, first beginning over two millennia ago in ancient Greece.

Although Hippocrates, the Father of Modern Medicine, and the Hippocratic corpus are some of our best and earliest sources of ancient Greek medical writing, they are not our earliest source of medical information. Rather than come from an ancient physician like Hippocrates, our earliest western source of medical information comes instead from the famous Greek poet Homer,[22] sometime between the 12th and 8th centuries BC. Although not a physician himself, Homer's epics the Iliad and the Odyssey provide a glimpse of medical ideas and practices centuries before Hippocrates through his depictions of Machaon and the actions he takes as a healer.[23] Furthermore, Homer's works introduce us to the ancient Greek's complex relationship between the divine and the physical in regards to disease. Although the plague at the beginning of Homer's Iliad is sent by Apollo, who is also ultimately the one to cure it,[24] and the Greek people are portrayed as understanding that something of this magnitude must have a divine origin – a sentiment echoed throughout ancient Greek and later early Christian history – yet the Gods are not considered the universal cause of all diseases. With explanations for symptoms and diseases ranging from a snake bite in Homer's Iliad, to Hesiod's depiction of diseases as spontaneous beings of their own,[25] we have information enough to see that not everything was attributed directly, or even indirectly, to the Gods. However, when the ancient Greeks did conclude that a disease or disaster was from the Gods, their explanation for it often echoed what many people throughout history thought: that people were being punished for angering a divinity – an explanation that we sometimes find in use even in modernity. Lastly, beyond all he does for providing us with an understanding of pre-Hippocratic medicine and disease, Homer also demonstrates through

[21] Vivian Nutton, Ancient Medicine (London: Routledge, 2004), 311.
[22] Nutton, Ancient Medicine, 37.
[23] Ibid.
[24] Ibid., 38-39.
[25] Ibid., 39-40.

his works that it was not only physicians who had access to this kind of medical information, as Homer himself had no issues presenting characters like Machaon, who embodied, amongst other things, the idealized physician. However, between Homer's contributions to our understanding of ancient medicine and the beginning of Hippocrates' career is an approximately four to eight hundred year gap, with little of anything that contributes to our understanding besides a part in Arctinus' Sack of Troy, where Machaon's brother, Podalirius, is made out to be as medically proficient as Machaon.[26] Now, while this addition, as well as the aforementioned description of diseases from Hesiod, are welcome additions to our knowledge base of ancient medicine, ultimately, they are additions that do little more than reinforce or restate knowledge already gleaned from older sources. As such, our story goes silent until Hippocrates, the Father of Modern Medicine, takes the stage centuries after Homer.

Unlike Homer, we have a solid idea of around what time Hippocrates was alive during, as well as when the Hippocratic medical sect began to spread, with Hippocrates being alive from the 5th to 4th centuries BC, and his sect beginning to spread during his lifetime. However, despite all that is attributed to Hippocrates, there's a great deal of uncertainty as to what works were written by Hippocrates himself, and which were written by his followers.[27] The surviving works that we have today from the Hippocratic tradition make up a collection called the Hippocratic Corpus. The Corpus in the form that we have it comprises around 60 pieces of work; however, the exact number of works that are in it, as well as the number of works that were once in it, are both open to debate.[28] On top of the uncertainty regarding the numbers of works, the authors of many of the works are also unknown, and the number of surviving texts within the Corpus that were personally

[26] Ibid., 39.
[27] "Hippocrates and the Hippocratic Corpus," University of Virginia, 2007, http://exhibits.hsl.virginia.edu/antiqua/humoral/.

authored by Hippocrates, if any were, is just as unclear. Despite the great fame attributed to Hippocrates, we know next to nothing about the man; it's not even clear whether or not Hippocrates himself devised the Hippocratic Oath.[29] However, it is believed that much of it was written around the period Hippocrates was alive and active, and, despite the great deal of uncertainty surrounding Hippocrates and much of the Corpus, regardless as to how much was written by him personally, it is more than clear that the impact that he and his Corpus had on the practice of medicine cannot be understated.

Hippocrates' main medical theory follows off of the philosophizing of the 5th century philosopher Empedocles and his theory that the world was composed of 4 elements.[30] From here, the Hippocratics explained how health and the human body fit into this theory. The Hippocratics argued that one's health depended upon keeping 4 basic fluids, or humors, in balance. Although the Hippocratic Corpus did not have a set, agreed upon number of significant humors,[31] the work On the Nature of Man discusses the 4 most common humors: blood, phlegm, yellow bile, and black bile,[32] the four of which are the ones that are most likely to be thought of even today when discussing humoural theory. Each of the humors corresponded with one of the 4 elements; air corresponded to blood, water to phlegm, earth to black bile, and fire to yellow bile.[33] This correspondence with the elements was the basis of the most common Hippocratic humoural theory - as well as one of the most enduring Hippocratic theories, lasting into modernity.[34] This set the groundwork for many theoretical explanations for diseases and sickness, as Hippocratics emphasized their explanations and treatments on the need to keep the humors, representative of the basic elements, in balance within the human body, and it was when these humors were displaced and imbalanced that we would see people become sick, with different diseases occurring depending

[28] Nutton, Ancient Medicine, 60.
[29] Ibid., 53.
[30] "Introduction to Late Medieval Medicine," Michigan State University, https://history.msu.edu/hst425/resources/online-essays/introduction-to-late-medieval-medicine/
[31] Nutton, Ancient Medicine, 79.
[32] Nancy Siraisi, Medieval and Early Renaissance Medicine: An Introduction to Knowledge and Practice (Chicago: University of Chicago Press, 2009), 104-105.
[33] "Introduction to Late Medieval Medicine," Michigan State University, https://history.msu.edu/hst425/resources/online-essays/introduction-to-late-medieval-medicine/.

on which humors were out of balance, as well as where they were out of balance.

Although Hippocrates is now by far the most famous physician of his time, the Hippocratics were not the only ones attempting to explain human sickness and health, and even amongst themselves there was not a universal consensus. Amongst the followers of Hippocrates, and spread throughout the Corpus, are a number of different interpretations of the same, limited observational information that was available at the time. From these shared observations of disease and how the body reacted to it came explanations for disease ranging from issues concerning the inability to properly digest food to the arguments from the author of Breaths of how all diseases are caused by air.[35] Along with humoural theory, some authors of the Corpus, like the author of The Sacred Disease, held that diseases were caused by the blockage of certain passages in the body, and that these blockages resulted in the body attempting to clear them out as could be seen in a person with a runny nose and teary eyes; for others, however, the inverse was held to be true, as some authors argued that these purges are not symptoms of a disease or the bodies attempts to cure itself, but instead that the purges themselves are the cause of the disease, so the fewer one has the healthier.[36] Similarly, amongst those physicians who believed diseases were caused by residues left over in the stomach due to incomplete digestion, the exact nature of the residues was disagreed upon. For Euryphon of Cnidus, residues were caused by an inability of the body to rid itself of all the food ingested, yet for his contemporary, Hedicus of Cnidus, the origin of residues had nothing to do with the bodies failure to dispose of extra food, but instead rested with a lack of exercise resulting in the food remaining undigested.[37] Further still, some authors from shortly after Hippocrates disagreed upon the exact relationship the humors had with disease, like Petron of

[34] "The Legacy of Humoral Medicine," AMA Journal of Ethics, Jul., 2002, https://journalofethics.ama-assn.org/article/legacy-humoral-medicine/2002-07.

[35] Nutton, Ancient Medicine, 73-74.

[36] Ibid., 73.

Aegina who argued that bile was only produced when a person was diseased, and that bile was not itself the cause of any diseases – a position directly opposed to the Hippocratic humoural theory – while others had separate but theoretically similar theories, like Philistion of Locri and his belief on three general causes of disease[38] – internal causes, which related to excesses or deficiencies in the four forms (opposed to the four humors); external causes, which related to wounds, sores, inadequate external heat or cold, and insufficient food; and the third cause, which was related to a failure of air to properly flow in and out of the body.[39]

Despite these differences in theories and explanations, as well as others left still undiscussed, all the Hippocratic physicians shared in common the view of and approach to medicine as a natural, physical phenomenon independent of divinity, and able to be explained and treated through logic and understanding. Although ancient physicians emphasized the significance of a variety of things that we know now are usually unrelated – like the emphasis from the author of Regimen, who stressed the importance of meteorological, geographical, and climatological factors when diagnosing and treating patients[40] – they did so under the reasonable belief that humans existed as a part of the greater world around us, and it was this same belief that led to the Hippocratic dismissal of the needless inclusion of divinity in favor of a logical, structured approach to medicine.[41] In general, Hippocratic physicians took a holistic approach to medicine regardless of the exact nature of their medical theories, and attempted to treat patients allopathically. Interestingly, because of the shared belief that health was dependent on the keeping and restoring of an individual's natural balance, whether that entailed humors or not, many treatments across theoretical divisions revolved around strict regimes of dieting, exercise, and managing bodily functions and affairs like

[37] Ibid.
[38] Ibid., 72-73.
[39] Ibid., 115.
[40] Ibid., 75.

sweating and excretion, making these the main focus of treatments for centuries to come, being far more common treatments across the board than the occasional use of therapeutic venesection, simple surgeries, and herbal medicines also seen at the time.[42] Regardless as to how much Hippocrates himself personally contributed to his Corpus, and how much of it was work done by contemporaries and followers of his that was later mistakenly attributed to him, the impact Hippocrates had through the dismissal of traditional beliefs and construction of causal accounts of health and disease independent of the mythological or magical explanations of the time can never be understated.[43] Surprisingly, the life and work of Hippocrates was not the only important event for ancient medicine happening in ancient Greece during the 5th century. Near the end of 5th century BC, a new healing God was introduced to Athens, and with it came a new spiritual cult of healing: the cult of Asclepius. However, rather than signifying an attempt to return to the traditional mythological explanations of disease and medicine that had begun to be replaced by natural, logical theories, the cult of Asclepius marked an interesting point of interplay between spiritual and conventional healing that would be echoed on and off throughout history. The earliest stories we have about Asclepius come from Homer in the Iliad, where he is the father of the previously discussed warrior healers Machaon and Podalirius. Like his sons after him, Asclepius was described as a renowned secular healer, with some modern scholars speculating that it was this skill as a healer that got Asclepius his divine status to begin with.[44] Although the healing performed by the cult itself was rather characteristic of the traditional spiritual healing methods of the time - performing a sacrifice and rituals before hoping for a visit from the God in their dreams so that they may be cured[45] an important distinction between the cult of Asclepius and the ones that came before it was Asclepius' intimate connec-

[41] Ibid., 57.
[42] Siraisi, Medieval and Early Renaissance Medicine, 1-2.
[43] Ibid.

tion with conventional medicine, given his background as a physician himself. Furthermore, although the initial method of treatment by the cult was irrefutably spiritual in nature, some accounts of the visions experienced by those visiting the cult of Asclepius for healing involved the God himself appearing and acting as a physician or surgeon in order to heal them, with many of the other treatments provided by the cult mirroring those done in contemporary medicine.[46] By the 3rd century BC it was already considered an 'ancestral custom' for physicians to make sacrifices to Asclepius biannually,[47] and looking back at the criticisms of spiritual healing present in the Hippocratic Corpus from the author of The Sacred Disease, it becomes easy to see why Hippocratic physicians could so easily accept a spiritual cult that arose in Hippocrates' lifetime; far from being atheistic attacks on the Greek spirituality of the time – as the author has no issues with the practices of praying and sacrificing to the Gods, or with those individuals who serve or choose to seek healing at temples – the author instead attacks "those who claim to be able to drive out demons and compel the Gods to do their will by means of chants and charms," as, where in temples the decision of healing is ultimately left to the will of the Gods themselves, these individuals claim a personal relationship with the divine as well as the power to use this relationship to heal upon their command. Thanks to the strong connection to secular healing that Asclepius had, and the proper application of appropriately pious spiritual healing, the rise of the cult of Asclepius became an example of spiritual and secular healing cooperating and coexisting with benefit for the greater good, and Asclepius himself became an symbol for both the power of the Gods to heal people and save lives, as well as a symbol for medicine and physicians themselves.[48]

Following the rise of the cult of Asclepius, besides the contributions and additions of some post-Hippocratic

[44] Nutton, Ancient Medicine, 104
[45] Ibid., 109.
[46] Ibid.
[47] Ibid., 111.

physicians – such as the briefly aforementioned Philistion of Locri and Petron of Aegina – little happens for the ancient understandings of medicine and disease until roughly one and a half centuries after the birth of Hippocrates, when Greece conquered Egypt. Up until Alexander the Great conquered Egypt, and realistically speaking during this period and for some time after as well, any form of anatomy was considered quite taboo for the ancient Greeks, as there were religious laws against interfering with a dead body. However, despite this widespread Greek taboo, and how it lasted through ancient Greece long into Christianity centuries later, it was not one held by ancient Egyptians, who the Greeks had long since known performed operations on the deceased in order to mummify their remains.[49] This knowledge, combined with the prejudice against native Egyptians during the early years of Alexandria, led to a few ancient Greek physicians residing in Alexandria to have the opportunity to dissect Egyptian corpses. The first to carry out this search into human anatomy was Herophilus of Chalcedon, who performed anatomical studies of the whole body and gave names to parts of the body, possibly including the Pineal Gland. Herophilus' studies led to him distinguishing between veins and arteries, differentiating motor and sensory nerves and examining their relationship with muscles during voluntary movement, and provided an explanation for pulsation, to name a few of his more significant feats.[50] Following after Herophilus was his contemporary Erasistratus of Ceos, a figure more controversial than his predecessor due to his tendency to ignore Hippocratic theories and to share challenging conclusions about the human body that did not match the conclusions of those who came before him, all of which led to later figures in medical history not giving his work the same regard as his predecessor's.[51] Despite this controversy though, Erasistratus contributed to the ancient anatomical under-

[48] Ibid., 112-114.
[49] Ibid., 129.
[50] Ibid., 132-133.

standing, potentially being the first to identify, examine, and describe all the valves in the heart and their workings, as well as investigating the human brain and concluding that the brain was the point of origin for nerves.[52] Additionally, unlike the static view of the human body presented by Herophilus, Erasistratus described the body as a living and functioning machine, explaining the functions of many body parts through mechanical analogies, and explaining many diseases as being the result of mechanical failure within the body, opposed to Hippocratic view of diseases common at the time.[53] Unfortunately, although Herophilus and Erasistratus marked the beginning of ancient Greek human anatomy, they also marked the end of it, as anatomical experimentation, as far as can be told, seems to have died out before the end of the 3rd century BC,[54] and was not to be resumed again until centuries later where debates about the morality and necessity of it continued.

After the anatomical developments of Herophilus and Erasistratus, and the subsequent pharmacological and surgical advancements made shortly later in the Hellenistic period,[55] the next large advancement for the understanding of medicine and disease came with the Roman conquest of Greece, as the following assimilation of Greek culture and medicine led to it becoming the foundation of western medicine, with some Greek medical theories continuing to be studied and defended well into the 19th century.[56] However, despite the significance of this transplantation, early on in the process the Roman people, many of whom were already opposed to anything foreign, were turned off the Greek import of medicine due to the poor and lasting reputation left in Rome by Archagathus, the supposed 'first doctor.' From there, many Romans didn't trust Greek physicians or medicine, a matter made worse by figures like Elder Cato who warned "against the corrupting influence of the Greeks, and especially their doctors."[57]

[51] Ibid., 133-137.
[52] "Erasistratus," Encyclopedia, Jul. 9, 2020, https://www.encyclopedia.com/people/medicine/medicine-biographies/erasistratus
[53] Ibid.
[54] Nutton, Ancient Medicine, 128.
[55] Ibid., 141-142.

Greek medicine was not kept down though, and by the mid 1st century,[58] though it was still considered not to be a practice for a proper Roman, it was commonplace for Greek physicians to be employed.[59] Despite the fact that even at this point in time Greek medicine was viewed as something inherently immoral, people were more than willing to reap the benefits of having Greek physicians and medicine around. A marker of when this attitude towards Greek medicine truly began to turn around was the success of the immigrant medical practitioner Asclepiades of Bithynia, a physician from around the turn of the millennium so successful that he was acknowledged across Rome. Along with gaining a significant reputation through his treatments, many of which notoriously involved wine and gentle exercise, Asclepiades also brought new theories along with him which, although they never overthrew the lasting impact of Hippocrates and his corpus, introduced a brief period where materialistic and mechanistic theories saw the light.[60] Regardless of the fact that his theories never had the same lasting effect of those of the Hippocratics, Asclepiades' work and success in Rome marked the true assimilation of Greek medicine into Roman society.

The next cornerstone of medical history comes two centuries after Asclepiades with the introduction of Galen, who took the work of Hippocrates and, in addition to making considerable contributions of his own to the field, re-shaped it in his own image,[61] revitalizing Hippocrates humoural theory and beginning Galenism, which would be the dominant form of medicine for over a millennium after Galen himself.[62] By the time Galen was practicing during the mid 1st century AD, surgeons had successfully been able to treat soldiers who had their stomachs cut open – so long as the intestines were undamaged – and by the peak of his career Galen himself expected any competent surgeon to be able to perform basic eye surgery, as well as be able to

[56] Ibid., 157.
[57] Ibid., 161-162.
[58] Interestingly, it was also around this period of time that St. Luke, author of the Gospel of Luke and of the Acts of the Apostles, lived, and he was traditionally held to be a Greek physician himself.
[59] Nutton, Ancient Medicine, 164-165.
[60] Ibid., 167-168.

treat a wide range of other issues such as ulcers, hernias, and aneurysms.[63] What first got Galen recognized, however, was his work as a surgeon in Pergamon, where the high priest retained him so that he could work on the priest's Gladiators.[64] There Galen developed quite a reputation for himself and received valuable practical and anatomical knowledge through treating the Gladiators, an opportunity not many physicians of the time had given the lasting taboo against dissection and anatomical research. After 5 years of this, Galen set off for Rome, where he went to work making a name for himself by disputing with the leading physicians there.[65] Beyond regularly disputing with authorities on subjects and writing works that called out other physicians or otherwise highlighted his position and interpretation of Hippocrates, Galen also furthered his reputation and the field through his advocacy and regular practice of anatomy, performing daily dissections, or sometimes vivisections, on what animals he could. Although extrapolating from animal dissection to human anatomy sometimes led Galen astray, his dissections led to remarkably accurate descriptions of the heart and the vascular system, many bones in the body, and even nerves and muscles – the latter being particularly significant given the lack of any modern aids when investigating some structures that are difficult to be seen by the naked eye – just to name a few, all the while improving his proficiency as a surgeon due to his familiarity with the body and his improved dexterity from all the practice.[66] Beyond his anatomical discoveries, Galen also furthered medicine through his medical theories, arguing that, based on his anatomical research, the human body consisted of three parallel systems originating with the liver, heart, and brain in order to serve different purposes in the body, which gave physicians a new way to understand how the human body functioned and how people could be treated and why those treatments work.[67] Due to both Galen's practical and theoretical work across his life, the

[61] Siraisi, Medieval and Early Renaissance Medicine, 4.
[62] Nutton, Ancient Medicine, 292.
[63] Ibid., 183.
[64] Ibid., 223.
Ibid., 224.

field of medicine had the groundwork for a unified understanding of the human body and how physicians could treat it, all leading to the point where, by 650 AD, learned physicians and intellectuals all 'knew' that the human body was organized anatomically into three systems, and that our health depended on the balance of humors which could change based on age and time of year, and the medical debates of ancient Greece and Rome had been replaced with debates on how best to interpret Galenism.[68] Galen's impact was so great that, when talking about conventional medicine, Galen's work would be the topic for the rest of the scope of this chapter, as Galenism stretched into modernity. However, while Galenism was becoming and as it was the dominant form of conventional medicine, equally significant developments were being made on the side of religious and spiritual healing and continued to be made well past Galen's lifetime as Christianity shaped the world.

Even before Constantine legalized Christianity and gave the Roman Empire the freedom of worship in 313 AD, the persecuted Christian community was making an impact on the well being of the sick and the poor,[69] if not yet in the field of medicine itself. Buildings called Diaconia were built as early as the second century AD, serving as a sort of early Hospital, where the sick and the poor were aided.[70] With the acceptance of Christianity within the Roman Empire, Christian communities became able to openly and actively organize the construction of Hospices and efforts to help those in need, and by 369 AD, St Basil of Caesarea had founded what is considered to be the first true hospital, a large scale, 300 bed building meant for the seriously ill and used to care for victims of the plague.[71] During this period of time, where Christianity quickly became the dominant religion, many religious leaders weren't particularly concerned with secular medicine, with most early Christians having a benign attitude towards medicine,

[66] Ibid., 232.
[67] Ibid., 232-234.
[68] Ibid., 292.

likely in part due to Galen's emphasis on there having been a benevolent Creator who put herbs and medicine in the world so that humanity could use them to relieve suffering – a notion easily compatible with belief in the Christian God.[72] In fact, there are a number of priests and even Bishops who continued to practice as physicians during this period, with many being renowned largely because of the medical skills they possessed. However, during the 4th and 5th centuries, medical knowledge was at times used as a weapon against those higher up in the religious community who possessed it, with Greek learning at times being compared to magic and heresy. As the attachment of medicine and healing to the old Gods drew the concern and suspicion from religious leaders, pagan healing shrines were destroyed and slowly replaced, a process which took place over multiple centuries.[73] Over time though, accounts of healing miracles began to abound, and the medicines of Christianity and Galen were soon considered complementary, as medicinal herbs and skills came from God and as such were valuable, but a reliance on them without the aid of prayer and faith was considered foolish.[74] It wouldn't take long before physicians, like most everyone else in the Christian world, would become Christians too, and from there, as the Christian world developed, so too did the practice of medicine. As the centuries passed, though it would be many centuries more before the hold of Galenism would be shook off or even questioned, developments were made in the practice of medicine. There was the 'Twelfth-century Renaissance,' leading to a growth in medical study,[75] the reintroduction of anatomy to the field in the thirteenth century when Venice decreed it mandatory practice for surgeons,[76] the first form of regulations for physicians and surgeons during the thirteenth and fourteenth centuries,[77] and restrictions from the largely instructional dissections being lifted in the fifteenth century.[78] Despite the many systemic and organizational developments made by the

---

[70] "History of Hospitals," Encyclopedia, Jun. 28, 2020, https www.encyclopedia.com/religio encyclopedias-almanacs-tra scripts-and-maps/hospitals-histo

[71] "The Christian Contributi to Medicine," Christian Medic Fellowship, 2000, https://ww cmf.org.uk/resources/publication content/?context=article&id=82

[72] Nutton, Ancient Medicin 302-303.

[73]. Ibid., 303-304.

Christian world leading up to the end of the medieval period, the number of actual medical advancements were few, with "certain basic physiological concepts and associated therapeutic methods - notably humoural theory and the practice of bloodletting to get rid of bad humors – [having] had a continuous life extending from Greek antiquity into the nineteenth century."[79] When the medieval period comes to a close near the end of the 15th century, though the world around it and the structures of the field itself look different, the actual practice of medicine is largely in the same place as it was nearly a millennium ago. Although the rise of Christianity did much to shape the medieval world and the structures therein, it wouldn't be until after that period ends, with the spread of syphilis running rampant in the 1490s, when the conversation about the possibility of diseases not explainable by complexional and humoural issues begins,[80] marking an important step towards our understanding of medicine as we have it today. But that and the events following are beyond the scope of this chapter.

When the medieval period ended, the practice of medicine had been around for several millennia, if not longer. Though it had changed greatly over its lifetime, being influenced by the events of the world around it, medicine still had a long way to go before it would wholly and truly cast off the views of antiquity. By the end of the 15th century, the world in which medicine then resided, much like the field of medicine itself, had been irrevocably changed, and it would only be a matter of time before our understanding of medicine and disease would advance along with the Christian world to where we are now.

[74] Ibid., 306.
[75] Siraisi, Medieval and Early Renaissance Medicine, 13-14.
[76] Ibid., 86-88.
[77] Ibid., 18-19.
[78] Ibid., 88-89.
[79] Ibid., 97.
[80] Ibid., 129.

IV.
THE CHRISTIAN VOCATION TO LOVE

> ❝
> DEAR FRIENDS, LET US LOVE ONE ANOTHER, FOR LOVE COMES FROM GOD. EVERYONE WHO LOVES HAS BEEN BORN OF GOD AND KNOWS GOD. WHOEVER DOES NOT LOVE DOES NOT KNOW GOD.[81]
>
> ❝
> GOD IS LOVE. WHOEVER LIVES IN LOVE LIVES IN GOD, AND GOD IN THEM.[82]

Before we begin a discussion on the Christian vocation to love, we first ought to understand its origins. This chapter will provide a brief overview on the fundamental ideologies regarding the Christian vision of the human condition, particularly through the Church's doctrine of the Blessed Trinity, and various passages within Sacred Scripture. As we will see in this chapter, it is ultimately humanity's shared relational existence within the Trinitarian nature of God that cultivates the Christian vocation to love and serve others in the world – especially those who are forgotten, living in poverty, or suffering from illnesses.[83] From a Christian anthropological perspective, humanity is born out of God's relational existence as Love.[84] In the Church, this relational existence is often referred to as the Trinitarian nature of God, or the Blessed Trinity. The presence of the Father, the Son, and the Holy Spirit, are interpreted as being distinct persons in a unified whole, that is God.

This paradoxical relationship of God existing as both three and one is the greatest foundation for the Church's

---

[81] New International Version, I John 4:7-8.
[82] I John 4:16.
[83] This is seen clearly in Matthew 25.
[84] The capitalized "L" refers to the perfect form of love, whereas the lowercase indicates a relational existence to this perfect form. This grammatical structure is commonly ascribed to Platonic metaphysics. It typically indicates the universal objectivity of a perfect, unchanging being or thing.

interpretation that God is Love. The union of these seemingly conflicting perspectives provides the foundation for a unity-in-difference and difference-in-unity that is crucial for the existence of Love. As Carlos Kloppenburg describes in Ecclesiology of Vatican II, the mission of the Church ought to reflect the invisible dimension of the Blessed Trinity in the corporeal world. Kloppenburg states that "[the] unification of the [Church] is a prolongation of the unity between the three divine Persons; by its own unity the people [share] in that other Unity, so that the unity of the Church cannot be understood apart from the Trinity."[85] Therefore, the Christian vocation to love, as the Church proposes, proceeds from the mysterious and paradoxical nature of the Blessed Trinity, as it is the Trinitarian nature of God that reveals His existence as Love.[86] Theologians, such as French Jesuit Henri de Lubac, place a great emphasis upon the simultaneity of the visible and invisible elements of the Church, and the importance of not separating them. For de Lubac, The mysterious relational existence of the Blessed Trinity serves as a model for how the Church ought to be in relationship with others. The sacraments that the Church provides are "[a] means of salvation [because] they are instruments of unity."[87] This striving towards unity emanates from the relational union revealed in the Trinitarian model of God. It is this unity-in-difference and difference-in-unity that reveals to many Christians the existence of God as Love.

Now, why is this the case? St. Bonaventure, a medieval Franciscan theologian, describes the significance of distinct persons within the Trinity as being fundamental for the existence of Love. He states that for God to exist as Love, there must exist an "other" to love. It is important that there exists a distinct "other", for if this were not the case, then the Father would be expressing a selfishness that is foreign to the very nature of love. It is the existence of differences or an "other" that allows for the possibility of

---

[85] Kloppenburg, Bonaventure. The Ecclesiology of Vatican II. Franciscan Herald Press, 1974, p. 25.

[86] I will be using masculine pronouns when discussing aspects of God as seen in various Church traditions; however, it is commonly held knowledge that God is beyond human gender or sex. The reason for my use of masculine pronouns is also due to the fact that many - if not all - of the explanations of the Blessed Trinity use the masculine form of God. It is important to note, however, that both female and male images of God are found in Sacred Scripture.

[87] Doyle, Dennis M. Communion Ecclesiology: Visions and Versions. Orbis Books, 2000, p. 68.

love to exist and flourish. As St. Thomas Aquinas states in his Summa Theologiae: "To love is to will the good of the other."[88] Therefore, love requires the acceptance of someone as other. I will give a brief analogy to illustrate this: Let us say that I claim to love person X. If I say that I love person X only because he has the same beliefs, hobbies, interests, and values that I do, then perhaps I do not love person X. It would appear that I am merely loving myself through him. This self-oriented form of 'love' as shown in the analogy is in stark contrast to the self-giving love revealed in the Blessed Trinity. As revealed in the analogy, differences ought to exist for genuine self-giving love to exist. If I do not embrace the distinctiveness of person X, then my love for him would appear rather questionable. Therefore, for love to exist, there ought to be distinct differences between persons; thus, God is revealed as Love due to the distinct persons within the relational existence of the Blessed Trinity. Bonaventure's Trinitarian theology necessitates the existence of differences that are required in Love. The Father, the Son, and the Holy Spirit express three different modalities of love that are actualized within the relational existence of God. Therefore, God, the Father, and His Son, Jesus Christ, must be distinct from one another. Now, what role does the Holy Spirit play in this relational existence? We can understand the Spirit existing as the love between the Father and the Son. It is the presence of this third 'person' that depicts the fruitfulness of the relationship between them. This fruitfulness within the Blessed Trinity symbolizes the infinite fecundity that genuine love inevitably produces as a result of its self-giving nature.

It is the Spirit of love, cultivated from the Blessed Trinity, that generates a creation that is distinct from God. Therefore, creation, made up of distinct creatures and itself distinct from its creator, exists as united yet sepa-

[88] St. Thomas Aquinas, Summa Theologiae, I-II,26.

rate from God through the loving movement and power of the Spirit. The Spirit facilitates the paradoxical relationship of unity-in-difference and difference-in-unity that is necessary for God's existence as Love. This emphasis on relationality, interconnectedness, and love, is cultivated from the very roots of Christian revelation. In A Trinitarian Anthropology, Michele M. Schumacher states, referencing Adrienne von Speyr and Hans Urs von Balthasar, that in the history of salvation, God has revealed His relational existence as Love through His Spirit in the Incarnation of His Son in the visible, corporeal world. Therefore, human bodies are seen as "instrument[s] of obedience" for the loving union within the Father, Son, and Holy Spirit.[89] According to Church teaching, the body should always be an instrument of unity; it ought to be used to express the Christian mission to love and serve others in the world.

In Genesis 1:26-7, God is described as having created humankind in His image and likeness. This indicates that humanity contains both "[c]orporeal and immaterial dimensions" in its existence.[90] This balanced union of body and spirit is fundamental to Christian anthropology. For instance, if Christians place too great of an emphasis on the immaterial dimension of humanity, then they run the risk of fleeing from the world rather than living within it. Christians who focus solely on the spiritual dimension may attempt to avoid all things related to the corporeal, material world. This extreme form of asceticism would make helping those who are sick or struggling with poverty meaningless since what is corporeal i.e., the body or the physical environment, is to be avoided. On the other hand, if there is a greater focus on the corporeal dimension, then this may produce more naturalistic ideologies that may eventually reject concepts pertaining to what is immaterial i.e., the soul or spirit. If this occurs, then the idea of an objective morality could cease to exist since moral concepts i.e., what

[89] Schumacher, Michelle M. A Trinitarian Anthropology: Adrienne Von Speyr & Hans Urs Von Balthasar in Dialogue with Thomas Aquinas Washington, D.C.: Catholic University of America Press, 2014, p. 17

[90] Sachs, John Randall. 1991 The Christian vision of humanity: basic Christian anthropology Collegeville, MN: Liturgical Press p. 52-7.

is good, bad, right, or wrong, are not explicitly observable in the natural world; thus, a form of moral relativism or apathy could arise. As a consequence of a relativistic view of morality, individuals may begin asking themselves whether genuine selflessness is justified, or even possible. Similar to the asceticism mentioned previously, individuals might not feel the need to fulfill any moral obligations as a result of their skepticism towards any existing objectivity regarding moral concepts.

Therefore, a Christian anthropology ought to propose a balance between the corporeal and immaterial dimensions of the human condition. As John Sachs describes in The Christian Vision of Humanity, the vocation to love necessitates an intentional integration of oneself into the Love revealed in the Blessed Trinity.[91] To live in the Spirit of love is to not only cultivate a virtuous interior life, it also demands for visible service within one's community. As revealed to us in Sacred Scripture: "Love the Lord your God with all your heart and with all your soul and with all your mind. This is the first and greatest commandment. And the second is like it: 'Love your neighbor as yourself.' All the Law and the Prophets hang on these two commandments."[92] The Church and other Christian communities have responded to this vocation to love in various ways throughout history. In the later chapters, we will see how the Church and other Christians have acted in response to different pandemics.

However, before we look at the various ways in which the Christian vocation to love has been expressed throughout history, it is important that we address the elephant in the room: If God exists as Love, then how could He let His creation suffer and experience death? Can a loving God coexist with the reality of disease, famines, natural disasters, and plagues? Why would a loving God permit the existence of evil in His creation? This "Problem of

---

[91] Sachs, John Randall. The Christian vision of humanity: basic Christian anthropology. p. 71.
[92] Matthew 22:37-40.

Evil" has puzzled philosophers and theologians for many centuries.[93] It is difficult to reconcile the existence of evil and suffering in the world with the idea of a God who not only loves, but is Love. So, what are we to do?

> "WHY IS LIGHT GIVEN TO THOSE IN MISERY, AND LIFE TO THE BITTER OF SOUL, TO THOSE WHO LONG FOR DEATH THAT DOES NOT COME, WHO SEARCH FOR IT MORE THAN FOR HIDDEN TREASURE, WHO ARE FILLED WITH GLADNESS AND REJOICE WHEN THEY REACH THE GRAVE? WHY IS LIFE GIVEN TO A MAN WHOSE WAY IS HIDDEN, WHOM GOD HAS HEDGED IN? FOR SIGHING HAS BECOME MY DAILY FOOD; MY GROANS POUR OUT LIKE WATER. WHAT I FEARED HAS COME UPON ME; WHAT I DREADED HAS HAPPENED TO ME. I HAVE NO PEACE, NO QUIETNESS; I HAVE NO REST, BUT ONLY TURMOIL."[94]

There is no other book in the Bible that faces the problem of evil and suffering with greater power than the poignant and enigmatic story of Job. The Book of Job in Sacred Scripture deals with the question of suffering and attempts to reconcile the existence of a loving God with the many atrocities and disasters that individuals face in the world. As explained in the Fourth Lateran Council, however great the similarity between God and creatures, there is always an ever-greater dissimilarity between them.[95] The Book of Job is a reminder to Christians that God, as creator of all, is infinitely greater than humanity. The Book of Job depicts the mystery of human existence as something that ought to transcend, and ultimately, transform the meaning of evil and suffering in the world. God's response to Job indicates the human need for a deeper faithfulness and humility in Him.

[93] Tooley, Michael, "The Problem of Evil", The Stanford Encyclopedia of Philosophy (Spring 2019 Edition), Edward N. Zalta (ed.).

[94] 3 Job: 20-6.

[95] Leclercq, Henri. "Fourth Lateran Council (1215)." The Catholic Encyclopedia. Vol. 9. New York: Robert Appleton Company, 1910.

> *"HAVE YOU JOURNEYED TO THE SPRINGS OF THE SEA OR WALKED IN THE RECESSES OF THE DEEP? HAVE THE GATES OF DEATH BEEN SHOWN TO YOU? HAVE YOU SEEN THE GATES OF THE DEEPEST DARKNESS? HAVE YOU COMPREHENDED THE VAST EXPANSES OF THE EARTH? TELL ME, IF YOU KNOW ALL THIS. WHAT IS THE WAY TO THE ABODE OF LIGHT? AND WHERE DOES DARKNESS RESIDE? CAN YOU TAKE THEM TO THEIR PLACES? DO YOU KNOW THE PATHS TO THEIR DWELLINGS?"*[96]

The mystery of pain and suffering has a distinctive capacity to stir individuals towards a deeper contemplation of the world. Individuals can cultivate wisdom by reflecting on these mysteries, discovering how they ought to live in light of it. In his apostolic letter, Salvifici Doloris (On the Christian Meaning of Human Suffering), Pope John Paul II describes how the choice of faith enlarges the horizon of meaning surrounding human death and suffering. Instead of coming to the conclusion that suffering has no meaning beyond the bitterness that it yields, the choice of faith turns individuals outward to contemplate their pain in relation to a mysterious, transcendent reality. Although suffering remains for many Christians a dilemma, by connecting human suffering to purposes and meanings beyond themselves, they can allow their suffering to be used for the sake of others.

An argument could be made that if evil and suffering did not exist, then it may be the case that everyone would believe in God; however, this would undermine human freedom and autonomy because individuals would not have the ability to choose otherwise.[97] It is important that humanity has the choice and freedom to believe in God; because if this were not the case, then God could not exist as Love. If, for example, there was a large neon sign in the sky from God that affirmed His existence, then it would not be implausible to suggest that many individ-

---

[96] 38 Job: 16-20.
[97] The Christian perspective on autonomy and freedom will be discussed further in Chapter 13.
[98] This is with the assumption that individuals have a clear way of knowing that it was God who made this sign, and that they were not being deceived in any kind of way.

uals would believe in God due to this act.[98] However, if this act did take place, then human autonomy and freedom would be impaired or even non-existent. For God to exist as Love, humanity must have the freedom to accept or deny Him. Therefore, an argument could be made that the existence of evil and suffering in the world does not necessarily negate the existence of God as being Love or loving. As Dónal P. O'Mathúna states in *Christian Theology and Disasters*:

> "A ROBOT CAN BE PROGRAMMED TO ALWAYS OBEY ITS OWNER, BUT THEN THE RELATIONSHIP BETWEEN THE TWO WOULD NOT BE PERSONAL. FREEDOM RISKS PAIN, AND HENCE A CHILD CAN REJECT HIS PARENTS, A SPOUSE CAN BE UNFAITHFUL, OR A PARENT CAN BE ABUSIVE. THESE RISKS [APPEAR] NECESSARY IN A WORLD WHERE FREEDOM, LOVE AND PERSONAL RELATIONSHIPS EXIST."[99]

[99] O'Mathúna, Dónal. (2018). Christian Theology and Disasters: Where is God in All This? 10.1007/978-3-319-92722-0_3.

In this chapter, we have seen how the Christian vocation to love demands that individuals be accepting of others and their differences in order for genuine love to come to fruition. It is through the loving of one's neighbour that an individual praises and loves God. Therefore, Christians must strive to express their love of God through the love they have for one another. Acts of charity, forgiveness, humility, and selflessness all serve to display the loving movement of God's Spirit of Love working in and through individuals. As we will examine in the later chapters of this book, the Christian vocation to love has manifested itself in different ways throughout history. When facing various pandemics, the Christian vocation to love can become distorted, disregarded, or forgotten entirely; and sometimes, we are able to see how individuals have appropriately responded to the call to love despite the dire circumstances that they have found themselves in.

V.
THE BLACK DEATH

> ❝ Dear But for this calamity it is quite impossible either to express in words or to conceive in thought any explanation, except indeed to refer it to God. For it did not come in a part of the world nor upon certain men, nor did it confine itself to any season of the year, so that from such circumstances it might be possible to find subtle explanations of a cause, but it embraced the entire world, and blighted the lives of all men, though differing from one another in the most marked degree, respecting neither sex nor age."
> - Procopius of Caesarea, 540 CE[100]

[100] Procopius, History of the Wars, 7 Vols., trans. H. B. Dewing, Loeb Library of the Greek and Roman Classics, (Cambridge, Mass.: Harvard University Press, 1914), Vol. I, 451-473.

The Black Death was one of the most deadly and influential pandemics in human history. In the peak of the outbreak in Europe between 1347 and 1351, the Black Death was responsible for the eradication of somewhere between a quarter to one half of Europe's entire population. Estimates of the death toll of the Black Death in Europe vary between 20 and 100 million, but even a more conservative estimate of 25 million deaths means that in Europe, on average, almost 14,000 people died per day in the five years between 1347 and 1351. The Black Death is outstanding among historical pandemics primarily because of the unprecedented percentage of the world's population that was wiped out as a result. The peak period of the Spanish Flu, which will be discussed later in this book, lasted from 1918 to 1920 and had a death toll in Europe comparable to the Black Death's. At the time of the Spanish Flu, however, the population of the world was well over 1.2 billion, while the population of Europe before the Black Death may have been around 440 million.[101] By the end of the 14th century, the world population had fallen to around 350 million, mostly as a result of the Black Death but also due to other factors, such as widespread famine. It took somewhere around 150 years for the world population to return to pre-Black Death levels. This sheer drop in population is almost incomprehensible from a modern perspective; for that proportion of the population to disappear today, a billion and a half people would have to perish. As is easy to imagine, the Black Death had a massive impact on the course of European history, leading to huge shifts in the economic landscape, society, and culture of Europe. This chapter will provide a general overview of the Black Death, the historical context of Europe leading up to the outbreak, and some of the consequences of the pandemic. In addition, we will touch upon the presence of plague in the modern world, as well as other plague pandemics, both prior and subsequent to the Black Death.

The infectious disease responsible for the Black Death was plague, an affliction caused by the bacterium *Yersinia pestis*, or *Y. pestis* for short. It is important to note that while the term "plague" has often been used as a general term for an outbreak of a disease of any kind, in this chapter the more technically correct definition of plague as, "the disease caused by the transmission of *Y. pestis*," will be used. There are three different kinds of plague which can occur in a person, depending on the method of transmission: bubonic plague, pneumonic plague, and septicemic

[101] "Historical Estimates of Worl Population," United States Cens Bureau, July 28, 2018, https: www.census.gov/data/table time-series/demo/internatio al-programs/historical-est-worl pop.html.

plague. The most recognizable form of plague is bubonic, which is primarily spread from bites from infected fleas, which move from animal host to human or from human to human. Bubonic plague is characterized by the presence of "buboes," which are painful swellings of the lymph nodes which typically occur in the armpits, groin, and neck area. Pneumonic plague is an infection of the lungs caused from inhalation of plague bacteria, and is spread directly from person-to-person. Septicemic plague is the rarest and deadliest form of plague, which is an infection of the blood rather than the lymph nodes or lungs. The three forms of plague are not mutually exclusive, since a person may be infected with only one type initially with the infection eventually spreading throughout the body leading to one of the other forms of plague, or someone infected with bubonic plague may infect someone else with pneumonic or septicemic plague. Today, plague can be easily treated with antibiotics, as long as it is detected as early as possible. With early treatment, the mortality rate of plague can be lowered to around ten percent, versus 30-60% for untreated bubonic plague and nearly 100% for untreated pneumonic or septicemic plague.[102, 103] In the 14th century, nobody knew what plague was caused by, let alone how to treat it. Catching the plague during the Black Death and surviving wasn't a matter of whether or not you received treatment in time, it was just a matter of chance.

Before discussing the Black Death itself, it is worth noting that plague had visited Europe many times before. The Plague of Justinian was the first outbreak of the plague in Europe, and occurred just over 800 years years before the Black Death, between 541-549 AD. Significantly less is known about the Plague of Justinian than of the Black Death, resulting in uncertainty about the Plague of Justinian's historical significance and death toll as a result. It has been commonly held that the Plague of Justinian was essentially as deadly and impactful as the Black Death, killing tens of millions of people and resulting in massive

[102] "Frequently Asked Questions | Plague | CDC," Centers for Disease Control and Prevention, Nov. 26, 2019, https://www.cdc.gov/plague/faq/index.html.
[103] "Plague," World Health Organization, Oct. 31, 2017, https://www.who.int/en/news-room/factsheets/detail/plague.

shifts in late antiquarian society. Recent studies have offered a more revisionist perspective, asserting that there is insufficient evidence to support such a maximalist view of the Plague of Justinian.[104] Regardless of the larger-scale interpretation of the Plague, the outbreak had an undeniable impact in Constantinople, the largest and most important city in Europe at the time. Procopius, advisor to Emperor Justinian I (for whom the plague was named), describes the extent of the mortality in Constantinople:

> "AND WHEN IT CAME ABOUT THAT ALL THE TOMBS WHICH HAD EXISTED PREVIOUSLY WERE FILLED WITH THE DEAD, THEN THEY DUG UP ALL THE PLACES ABOUT THE CITY ONE AFTER THE OTHER, LAID THE DEAD THERE, EACH ONE AS HE COULD, AND DEPARTED; BUT LATER ON THOSE WHO WERE MAKING THESE TRENCHES, NO LONGER ABLE TO KEEP UP WITH THE NUMBER OF THE DYING, MOUNTED THE TOWERS OF THE FORTIFICATIONS IN SYCAE, AND TEARING OFF THE ROOFS THREW THE BODIES THERE IN COMPLETE DISORDER; AND THEY PILED THEM UP JUST AS EACH ONE HAPPENED TO FALL, AND FILLED PRACTICALLY ALL THE TOWERS WITH CORPSES, AND THEN COVERED THEM AGAIN WITH THEIR ROOFS. AS A RESULT OF THIS AN EVIL STENCH PERVADED THE CITY AND DISTRESSED THE INHABITANTS STILL MORE, AND ESPECIALLY WHENEVER THE WIND BLEW FRESH FROM THAT QUARTER."

It is estimated that around one-fifth of the population of Constantiniople died during the course of the plague. Similar descriptions of the plague would be seen in the time of the Black Death, and it is estimated that around one-fifth of the population of Constantinople passed away.[105] In a pattern that would be repeated with the Black Death, the plague returned in numerous smaller outbreaks until the 8th century, after which it disappeared from Europe for several centuries. The Plague of Justinian

[104] Lee Mordechai et al. "The Justinianic Plague: An inconsequential pandemic?" Proceedings of the National Academy of Sciences of the United States of America 116, no. 51 (2019): 25546-25554. doi:10.1073/pnas.1903797116

[105] "Plague, Justinianic (Early Medieval Pandemic)," in The Oxford Dictionary of Late Antiquity, ed. Oliver Nicholson (Oxford University Press, 2018), 1200-01.

and the subsequent outbreaks are referred to as the "first plague pandemic," while the Black Death and its subsequent outbreaks make up the "second plague pandemic." In the time between the two outbreaks, understanding of the plague's origins would not advance, and any immunity against the plague which may have built up in the population would eventually disappear, slowly setting the stage for the Black Death.

By the time plague arrived in Europe in 1347, the continent had already been enduring decades of widespread strife and hardship. Up until the end of the 13th century, Europeans had enjoyed what is now referred to as the "Medieval Warm Period," a time of unusually warm and climate in the North Atlantic region which occurred between the 10th and 13th centuries. With this warmer weather came an agricultural revolution in Europe, bringing new farming techniques, widespread clearing of forests and swamps to make way for more farmland, and flourishing trade and commerce. Peasant farmers (who made up a majority of the population at the time) enjoyed a somewhat higher standard of living, with better diets and longer life expectancies contributing to massive population growth. These prosperous times would not last, however. As the 13th century came to a close the Medieval Warm Period ended, with less favorable weather lowering food production across a now overpopulated Europe. In 1314, an extremely rainy autumn made the way for several years of wet winters and cold summers, leading to even further reductions in crop harvests. All this would culminate in The Great Famine of 1315-1317, which is known to be the worst famine in the history of Europe, resulting in the deaths of at least ten percent of the population.[106]

On top of the immense loss of life, the famine had devastated Europe's reservoir of livestock, and caused violence and crime to run rampant throughout the conti-

[106] Judish Bennett, Medieval Europe: A Short History, Eleventh Edition (New York: McGraw Hill, 2011), 303.

nent.[107] The final major crisis to strike Europe before plague arrived was the beginning of the Hundred Years' War between France and England in 1337. The Black Death arrived during a time of hardship in Europe, which would only serve to magnify the devastation the plague would leave in its wake.

The Black Death originated in central Asia, where it could be found hosted in fleas among the local rat population. In the fifteen years before the plague arrived on sailing ships in Constantinople (modern-day Istanbul), the Black Death ravaged across China, India, and the rest of Asia, killing somewhere around 25 million people.[108] Plague-carrying Italian trading ships left Crimea and crossed the Black Sea, passing through Constantinople (modern-day Istanbul) in the spring of 1347 on their way to the Mediterranean Sea.[109] By the end of the year, the plague reached the southern shores of France, and spread throughout the islands of Corsica, Sardinia, and Sicily. Over the next year, the plague continued to march northwards further into Europe, arriving in Paris in the summer of 1348. By 1350 virtually all of Europe was in the grip of the Black Death.

The Black Death left misery and terror wherever it went. First-hand witnesses such as friar Jean de Venette, renaissance humanist Giovanni Boccaccio, and chronicler Agnolo di Tura wrote descriptions of the plague's effects on the cities of Paris, Sienna, and Florence, respectively.[110] Common themes can be found across their accounts, such as the gruesome swellings caused by bubonic plague, the complete ineffectiveness of any plague treatment or preventative remedy, the necessity of mass graves to handle the numbers of bodies being brought to the city churches, the abandonment of the sick to die in their homes or in the street, and the overall disintegration of public order and social ties. The following account by Agnolo di Tura provides a particularly vivid and personal picture of the plague:

[109] John Aberth, The Black Death: The Great Mortality of 1348-1350 New York: Palgrave Macmillan, 2005), 13.

[110] Perry Rogers, Aspects of Western Civilization, (Prentice Hall, 2000), 353-65.

The mortality began in Siena in May (1348). It was a cruel and horrible thing; and I do not know where to begin to tell of the cruelty and the pitiless ways. It seemed to almost everyone that one became stupefied by seeing the pain. And it is impossible for the human tongue to recount the awful thing. Indeed one who did not see such horribleness can be called blessed. And the victims died almost immediately. They would swell beneath their armpits and in their groins, and fall over dead while talking. Father abandoned child, wife husband, one brother another; for this illness seemed to strike through the breath and sight. And so they died. And none could be found to bury the dead for money or friendship. Members of a household brought their dead to a ditch as best they could, without priest, without divine offices. Nor did the death bell sound. And in many places in Siena great pits were dug and piled deep with the multitude of dead. And they died by the hundreds both day and night, and all were thrown in those ditches and covered over with earth. And as soon as those ditches were filled more were dug.

[111] Rogers, Aspects, 353-65

"AND I, AGNOLO DI TURA, CALLED THE FAT, BURIED MY FIVE CHILDREN WITH MY OWN HANDS. AND THERE WERE ALSO THOSE WHO WERE SO SPARSELY COVERED WITH EARTH THAT THE DOGS DRAGGED THEM FORTH AND DEVOURED MANY BODIES THROUGHOUT THE CITY. THERE WAS NO ONE WHO WEPT FOR ANY DEATH, FOR ALL AWAITED DEATH. AND SO MANY DIED THAT ALL BELIEVED THAT IT WAS THE END OF THE WORLD. AND NO MEDICINE OR ANY OTHER DEFENSE AVAILED".[111]

A disaster this grotesque and vast in scale warranted some sort of an explanation, and though many attempted to explain the origins of plague, none even came close to the true causes. One of the most popular theories put forward was that the sickness was caused by a great miasma, a movement of "bad air" across Europe causing peoples' bodily humors to become unbalanced. Naturally, people imagined that if there's such a thing as bad air, there must be good air, and so a common preventative measure against the plague (and illness in general) was to adorn oneself with good-smelling herbs and flowers to keep the rotten air away. Another somewhat unpopular theory proposed by the University of Paris blamed the plague on an unfavorable alignment of celestial bodies.[112] These theories did little to change the reality of the situation in Europe, and in some places frustrated Christians instead blamed Jews for the pestilence, resulting in massacres of entire Jewish communities in some parts of Europe. Not everyone supported this narrative, such as German writer Kondad of Megenberg, who wrote that the "Jewish people are justly detested by us Christians in accordance with the fundamentals of the Catholic faith," but still in general opposed the scapegoating of the Jews. With no way of treating, explaining, or effectively containing the plague, the Black Death was able to blanket almost all of Europe freely. In its wake it left bodies, a shaken surviving populace, and a greatly changed social landscape. As the Black Death gradually slowed and eventually ended in 1351, Europe found itself in the middle of an agricultural labour shortage. The population decrease and abandonment of farms and villages meant that a dependable rural workforce was now much more important. Peasants working on their noble landowners' fields found their labour to have increased significantly in value, meaning they found themselves able to bargain for wage increases, improving their standard of living and heightening their positions in society. The aftermath of the Black Death also saw a decline in serfdom (not in eastern Europe), the system of indentured servitude under which many peasants lived. Many landowners attempted, successfully or unsuccessfully, to quell these advancements in the power of the peasants. The full societal effects of the Black Death are too complicated to fully discuss in this chapter, and

[112] Aberth, The Black Death, 38

many of the changes seen in the time after the Black Death had already been put in motion years before the plague had even reached Europe.

The second plague pandemic would continue until the early 19th century with intermittent outbreaks across Europe and the Middle East. None of the outbreaks would even approach the severity of the Black Death, with death tolls for individual epidemics mostly numbering in the tens or hundreds of thousands. Beyond the Black Death, the plague became less of an apocalyptic act of God and more of a source of general anxiety and fear. It was during one of these outbreaks in 1377 that the term "quarantine" was coined. In Ragusa (modern day Dubrovnik, Croatia), the city's council established a law requiring those visiting the city to remain in a *trentino*, a thirty-day period of isolation, which was eventually lengthened to a forty-day *quarantino*.[113] In addition, the modern-day image of a beak-masked cloak-wearing plague doctor comes from 17th and 18th century descriptions and illustrations. Numerous explanations have been proposed for why plague outbreaks declined in severity, including improved sanitation and hygiene, better nutrition, immunity developing over time, and improvements in quarantine protocols.[114] Even though the plague was still poorly understood, improvements in living standards and outbreak management may have helped ensure that a second Black Death never occurred.

The third plague pandemic began in 1855 in the Yunnan province of China, and would go on to be the first time that the plague infected every inhabited continent on Earth. By the time the plague arrived once again in Europe in 1899, medical knowledge and sanitation techniques had evolved significantly since the second plague pandemic, with germ theory becoming widely accepted. The same year, bacteriologists Kitasato Shibasaburō and Alexandre Yerstin both discovered, completely independently from

[113] Philip Mackoviak and Paul Sehdev, "The Origin of Quarantine," Clinical Infectious Diseases 35, no. 9, (Nov. 2002): 1071-1072, doi.org/10.1086/344062.

[114] Andrew B. Appleby,"The Disappearance of Plague: A Continuing Puzzle," The Economic History Review 33, no. 2, (1980): 161–173, www.jstor.org/stable/2595837

each other, the bacterium responsible for bubonic plague - with Shibasaburō discovering the bacterium only a few days before Yerstin. A year before, French physician Paul-Louis Simond published a paper detailing his discovery that fleas spread bubonic plague from rat to rat and from rat to human. Between 1899 and 1947, there were 1692 cases and 457 deaths due to the plague in Europe.[115] While this plague pandemic's effects on Europe were minimal, other parts of the world suffered greatly. This time, the plague would be deadliest in colonial British India, where between 1896 and 1921, it was responsible for the deaths of somewhere around 12 million Indians. Comparatively, in the rest of the world collectively about 3 million people died.[116] In Manchuria, 60,000 people died as a result of an outbreak of pneumonic plague between 1910 and 1911. The third plague pandemic ended in the 1950s. While this most recent plague pandemic demonstrates a great deal of progress in the management of plague outbreaks, especially looking at the relatively small number of deaths in Europe, this event still resulted in the loss of millions of humans and the worldwide spread of the *Y. pestis* bacterium. According to the World Health Organization, there were 3248 recorded plague cases and 584 deaths between 2010 and 2015. The countries where plague is most endemic are the Deomcratic Republic of the Congo, Madagascar, and Peru. Again, the mortality rate of any of the three kinds of plague can be dramatically lowered with antibiotic treatment. Today, plague is the most dangerous in developing countries where people may not have access to the early detection and treatment needed to manage the plague. Regardless of what period in time plague has occurred, the disease is always responsible for immense human suffering and death. Hopefully, in the future, plague may be remembered in the same way we now think of smallpox: as a relic of the past, and nothing more.

[115] Barbara Bramanti et al., "The Third Plague Pandemic in Europe," Proceedings. Biological sciences 286, no. 1901 (2019): 20182429. doi:10.1098/rspb.2018.2429

[116] Aanchal Malhotra, "When the 1897 bubonic plague ravaged India," Livemint, Apr. 26, 2020, https://www.livemint.com/mint-lounge/features/when-the-1897-bubonic-plague-ravaged-india-11587876174403.html.

VI.
THE BLACK DEATH AND CHRISTIANITY

> *In the midst of such illness, they alone [the Christians] showed their sympathy and humanity through their deeds. Every day some continued caring for and burying the dead, for there were multitudes who had no one to care for them; others collected those who were afflicted by the famine throughout the entire city into one place, and gave bread to them all."*
>
> *-Eusebius of Cesarea, ca. 314 A.D.*[117]

[117] Eusebius, Church History, 9.8.14 (adapted for readability).

Even just the word "plague" is equated with terrible suffering and pain, hence the phrase "avoid it like the plague". The only apparent remedy was to simply avoid it at all costs. As we have already seen in the preceding chapter, the specific outbreak of plague commonly referred to as the 'Black Death' was the cause of tremendous suffering for the entire population of Europe during the years 1347-1351 A.D. There was no one left untouched by the effects of the plague. Those who survived, if they had been so fortunate as to have either recovered or avoided catching it in the first place, would have all lost loved ones or people they knew to this devastating pestilence. In this chapter, the goal is to examine some of the effects the Black Death and other outbreaks of the plague have had on Christianity and the Church up and down the centuries. We'll begin by looking at the possibility that pandemics actually contributed to the rise of Christianity within the context of the primarily pagan Roman Empire during the first several centuries, essentially by providing Christians with the opportunity to "practice what they preached". Next, we'll examine the Black Death itself and the effect it had on the primarily "Christian" European society of the time, including the kinds of responses this particular religious milieu elicited. Finally, we'll see how both the Black Death and subsequent outbreaks of the plague managed to have a galvanizing effect on some dedicated Christians, and take a brief look at what sort of a legacy they left behind.

"IN A GROUNDBREAKING 1996 BOOK, THE RISE OF CHRISTIANITY, THE AMERICAN SOCIOLOGIST RODNEY STARK TURNED THE TOOLS OF HIS TRADE ON THE EARLY CHURCH WITH FASCINATING RESULTS. AMONG OTHER THINGS, HE HIGHLIGHTED THE ROLE THAT SUCH PANDEMICS—AND CHRISTIANS' RESPONSE TO THEM, WHICH DIFFERED FROM OTHER PEOPLE'S—PLAYED IN THE ULTIMATE CHRISTIANIZATION OF THE ROMAN EMPIRE."[118]

Recognizing that the rise of Christianity within the Roman Empire was the combination of a number of factors, what Stark proposes is that the role played by pandemics should not be overlooked. Without going into too much detail, the essence of Stark's argument is that: "had classical society not been disrupted and demoralized by these catastrophes, Christianity might never have become so dominant a faith."[119] When compared with classical paganism and Hellenistic philosophy, for example—being unable to provide either explanations or comfort when confronted by such an epidemic—Christianity offered both a more "satisfactory account of why these terrible times had fallen upon humanity, and it projected a hopeful, even enthusiastic, portrait of the future."[120] He also suggests:

[118] Stephen Bullivant, Catholicism in the time of Coronavirus (Park Ridge, IL: Word on Fire Catholic Ministries, 2020), 5-6.

[119] Rodney Stark, The Rise of Christianity: A Sociologist Reconsiders History (Princeton, NJ: Princeton University Press, 1996), 75.

> "CHRISTIAN VALUES OF LOVE AND CHARITY HAD, FROM THE BEGINNING, BEEN TRANSLATED INTO NORMS OF SOCIAL SERVICE AND COMMUNITY SOLIDARITY. WHEN DISASTERS STRUCK, THE CHRISTIANS WERE BETTER ABLE TO COPE, AND THIS RESULTED IN SUBSTANTIALLY HIGHER RATES OF SURVIVAL. THIS MEANT THAT IN THE AFTERMATH OF EACH EPIDEMIC, CHRISTIANS MADE UP A LARGER PERCENTAGE OF THE POPULATION EVEN WITHOUT NEW CONVERTS."[121]

Acknowledging that this undoubtedly would have created a shift in the social demographics of the time, Stark concludes the role of epidemics played in the rise of Christianity simply cannot be ignored. Not only that, but on a deeper level he also recognizes that "something distinctive (came) into the world with the development of Judeo-Christian thought: the linking of a highly social ethical code with religion."[122] This distinction not only set Christianity apart from paganism, but it meant that "at a time when all other faiths were called to question, Christianity offered explanation and comfort. Even more important, Christian doctrine provided a prescription for action. That is, the Christian way appeared to work."[123] Given the number of times widespread epidemics struck in the years leading up to the Plague of Justinian, it stands to reason that, provided Christians really did "practice what they preached", they could hardly but have had an effect on society at large. Especially when there is evidence that the most common pagan response to epidemics was to flee for their lives and shun contact with the sick.[124] For, as Tertullian put it: "It is our care of the helpless, our practice of loving kindness that brands us in the eyes of many of our opponents. 'Only look,' they say, 'look how they love one another!'" [125] With all that in mind, as we turn to examine the Black Death itself, one of the first things to recognize is that when it struck in approximately 1347 A.D., Euro-

[120] Ibid.
[121] Ibid., 74-75
[122] Ibid., 86
[123] Ibid., 82
[124] Ibid., 85.

pean society was no longer predominantly pagan but at the peak of 'Christendom', which might be described as follows:

> "LATIN EUROPE WAS UNITED IN THE CENTRAL MIDDLE AGES NOT ONLY BY SUCH TANGIBLE THINGS AS WOOL, PAPAL LEGATES AND LETTERS, BUT ALSO BY A NEW WAY OF UNDERSTANDING ITSELF. THE ELEVENTH-CENTURY REFORM MOVEMENTS STIRRED UP THE DEEPLY HELD FEELING THAT THE CHRISTIAN RELIGION WAS MORE IMPORTANT, MORE REAL, THAN THE OTHER SOCIAL GROUPINGS, SUCH AS REGIONS OR KINGDOMS, IN WHICH PEOPLE LIVED." [126]

Now, that did not necessarily mean that the vast majority of the population embraced Christian values in a whole-hearted way. Much like today, there were plenty of people who were 'Christian', if not only in name, then at least by association. Moral standards, even among clergy, were notoriously lax; and even though the Christian way of life may have been considered the 'best way to live', not everyone was very good at putting it into practice. In the words of G.K. Chesterton: ""The Christian ideal has not been tried and found wanting; it has been found difficult and left untried."[127] This was on clear display during the havoc wreaked by the Black Death. For example, when it swept across Europe, both doctors and priests were expected to be on the front lines. Doctors, obviously to care for people's physical wellbeing, while priests were charged with their spiritual care. If nothing else, priests were expected to administer last rites to a person judged to be fatally ill. When faced with the possibility of contracting the plague, however, many people, including priests, refused to minister to the dying out of fear for their own safety. In the words of Franciscan Friar Michele da Piazza:

> "BECAUSE OF THE SCALE OF THE MORTALITY, MANY MESSINESE LOOKED TO MAKE CONFESSION OF THEIR

[125] Tertullian, Apology 39, 1989 e

[126] Joseph H. Lynch and Phill C. Adamo, The Medieval Church: Brief History, 2nd ed. (New York NY: Routledge, 2014), 177.

[127] G.K. Chesterton, What's Wron with the World (New York, N Dodd, Mead and Company, 1910 48.

> SINS AND TO MAKE THEIR WILLS, BUT PRIESTS, JUDGES, AND NOTARIES REFUSED TO VISIT THEM, AND IF ANYONE DID VISIT THEIR HOUSES, WHETHER TO HEAR CONFESSION OR DRAW UP A WILL, THEY WERE SOON SURE TO DIE THEMSELVES ... CORPSES LAY UNATTENDED IN THEIR OWN HOMES. NO PRIESTS, SONS, FATHERS, OR KINSMEN DARED TO ENTER; INSTEAD THEY PAID PORTERS LARGE SUMS TO CARRY THE BODIES TO BURIAL."[128]

Faced with an overwhelming number of casualties, combined with a general shortage of priests—in addition to those who refused to do their duty, there were a great number of priests did die from the plague, potentially upwards of 40%, based on clergy registers in England[129] —Bishop Ralph Shrewbury of Bath and Wells in England came up with a radical solution. In a letter to the clergy of his diocese dated Jan. 10, 1349, he states:

> "THE CONTAGIOUS PESTILENCE, WHICH IS NOW SPREADING EVERYWHERE, HAS LEFT MANY PARISH CHURCHES AND OTHER BENEFICES IN OUR DIOCESE WITHOUT AN INCUMBENT, SO THAT THEIR INHABITANTS ARE BEREFT OF A PRIEST. AND BECAUSE PRIESTS CANNOT BE FOUND FOR LOVE OR MONEY TO TAKE ON THE RESPONSIBILITY FOR THOSE PLACES AND VISIT THE SICK AND ADMINISTER THE SACRAMENTS OF THE CHURCH TO THEM—PERHAPS BECAUSE THEY FEAR THAT THEY WILL CATCH THE DISEASE THEMSELVES—WE UNDERSTAND THAT MANY PEOPLE ARE DYING WITHOUT THE SACRAMENT OF PENANCE, BECAUSE THEY DO NOT KNOW WHAT THEY OUGHT TO DO IN SUCH AN EMERGENCY AND BELIEVE THAT EVEN IN AN EMERGENCY, CONFESSION OF THEIR SINS IS OF NO USE OR WORTHLESS UNLESS MADE TO A PRIEST HAVING THE POWER OF THE KEYS. THEREFORE, DESIROUS AS WE MUST BE TO PROVIDE FOR THE SALVATION OF SOULS AND TO CALL BACK THE WANDERERS

---

[128] John Aberth, From the Brink of the Apocalypse: Confronting Famine, War, Plague, and Death in the Later Middle Ages, 2nd ed. (New York, NY: Routledge, 2010), 120.

[129] Ibid., 91.

> WHO HAVE STRAYED FROM THE WAY, WE ORDER AND FIRMLY ENJOIN YOU, UPON YOUR OBEDIENCE, TO MAKE IT KNOWN SPEEDILY AND PUBLICLY TO EVERYBODY, BUT PARTICULARLY TO THOSE WHO HAVE ALREADY FALLEN SICK, THAT IF ON THE POINT OF DEATH THEY CANNOT SECURE THE SERVICES OF A PROPERLY ORDAINED PRIEST, THEY SHOULD MAKE CONFESSION OF THEIR SINS, ACCORDING TO THE TEACHING OF THE APOSTLE, TO ANY LAY PERSON, EVEN TO A WOMAN, IF A MAN IS NOT AVAILABLE."[130]

This extraordinary provision was confirmed by Pope Clement VI in a papal indulgence a few months later, demonstrating just how serious the situation had become. Not only that, but in order to replace the sheer number of priests who had succumbed to the pestilence, there was a significant surge in ordinations in the years immediately following the Black Death. As more men were ordained to the priesthood, it was somewhat inevitable that many would have been seriously under-qualified, unfortunately contributing to the problem of low moral standards among clergy; however, it also indicates how important the sacraments were to people, and even amidst widespread criticism of priests in general, their 'services' were still in high demand. Even in the midst of a pandemic people in Christian Europe held on to their religious convictions, for unlike paganism during the Roman Empire, Christian doctrines on sin, death, heaven and hell, for example, not only provided a potential explanation for what was taking place, but also offered the possibility of comfort for those afflicted by the plague.

Similar to responses to the first outbreaks of plague around 541-750 A.D., the most common explanation for the devastation brought about by the Black Death was divine retribution. This notion likely originated due to scriptural accounts of famines and pestilences being the result of God's judgement or punishment for a particular

[130] Ibid., 121.

sin. Think of the story in Exodus 7-12 when God afflicts the Egyptians with a succession of ten 'plagues' (one of which included painful boils not unlike the bubonic plague), or in 2 Samuel 24 and the parallel account in 1 Chronicles 21 when, as retribution for taking a census, which was explicitly forbidden, King David is given the choice between famine, pestilence, or being subject to his enemies a period of time. If, therefore, the plague was the result of sinful behavior, true contrition and sincere repentance were considered the best possible remedies. For example, in his allegorical poem *On the Judgment of Sol at the Feasts of Saturn*, the Liège doctor Simon of Corvino "tells his audience that before taking doctor's medicines, they should first medicate the soul with 'holy remedies' such as a 'contrite heart' and 'tearful prayers.'" Doctors in general "urged their patients to tend to their spiritual well-being in the belief that this would created the best mental state for preservation and healing, as they then 'fear death less.'"[131] It was this sort of thinking that led to the so-called 'Flagellant movement', possibly the most infamous response to the Black Death. For even though the practice of prescribed penitential whippings dates back to St. Augustine in the 5th century, and St. Benedict in the 6th; in addition to the fact that there are records of processions similar to those of the Flagellants taking place prior to 1348, the Flagellant Movement became a point of contention because it was considered particularly radical in nature.[132] The main rationale behind the movement was "excessive atonement for excessive sin",[133] and therefore, believing the plague to be the result of sin, the extreme penance undertaken was believed to elicit mercy and forgiveness from God. Traveling from one town to another, the Flagellants would hold organized processions where they would whip themselves and pray: "that through such affliction of the body (namely their whippings), the epidemic disease would be evaded".[134] Even though they received much popular success, on Oct. 20, 1349, Pope

---

[131] Ibid., 120.

[132] Ibid., 136—In addition to their practice of whipping themselves, the Flagellant movement had incorporated a number of other elements, including a so-called "heavenly letter from Jerusalem" which supposedly prescribed the type of penance they had undertaken.

[133] Ibid., 137

Clement VI ordered the suppression of the Flagellant movement, calling it a "vain religion and superstitions invention". Though careful not to prohibit all penance, the pope insisted that it must be done "with a right intention and out of pure devotion…without the aforesaid superstitions, gatherings, associations, and conventicles" of the Flagellant movement.[135] his did not mean that the Flagellant movement suddenly disappeared overnight, however it did contribute to the movement gradually losing favor among the general population—especially given how the plague raged on, showing no signs of abating in response to their efforts.

Meanwhile, another popular explanation for the plague was the belief that the Jews were responsible for poisoning the wells. Originally taking root in southern France and Spain, widespread persecutions arose and some historians suggest that the association of the Jewish Rabbinical class with a esoteric Theosophy called 'Kabbalah'—potentially suspected as being akin to black magic—could have led to the suspicion that Jews were even capable of poisoning the water. A number of Church authorities and civic officials, including Pope Clement VI and King Peter of Aragon, sought to suppress violence against the Jews; however, throughout much of Europe such persecution persisted. Part of the reason this scapegoating of the Jews was considered justified is there had been a "contentious relationship between the Jewish and Christian communities" right from the beginning of Christianity:

[134] Ibid., 140
[135] Ibid., 145

> "THE CAUSES FOR THE TENSION INCLUDED SOCIAL, ECONOMIC, THEOLOGICAL, POLITICAL AND SPIRITUAL ONES. JEWS WERE VISIBLY OUTSIDERS, A SEPARATENESS EMPHASIZED BY THEIR DISTINCTIVE DRESS AND CUSTOMS, AND THE DIETARY AND RELIGIOUS PRACTICES DICTATED BY JEWISH LAW, AND THEIR CONTIN-

UED VISIBLE EXISTENCE, GIVEN THEIR REJECTION OF CHRIST, SEEMED AN INSULT TO THE TRUTH OF CHRISTIANTY."[136]

Unfortunately, this sort of unwarranted hostility persisted right up into the 20th century, as can be attested to by the tragic events associated with the Holocaust.[137] As these two examples have illustrated in a particularly vivid way, human nature instinctively seeks for meaning, especially when it comes to suffering, and people saw excessive atonement and scapegoating as two potential ways to find meaning in the midst of horrific circumstances. It must be noted, however, in no way does this condone their actions, but simply acknowledges part of the underlying motivation.

That all being said, even in the midst of the worst of tragedies, there are always a few examples of people who rose to the challenge despite extraordinary difficulty and hardship. A number of "careerist" prelates made serious reform efforts in the aftermath of the Black Death. Including John Thoresby, chancellor of England during the 1350's and archbishop of York until his death in 1373. In the post-plague years he devoted himself to pastoral work, such as recruiting candidates for ordination, compelling priests to return to their parishes, improving administration of the sacraments, as well as attempts to improve religious education for both the laity and clergy alike. Another prime example being St. Charles Boromeo:

> "FOR HIS PART, ST. CHARLES BORROMEO WAS THE PRIVILEGED SCION OF A MEDICI WHO WORRIED MORE ABOUT HIS APPEARANCE AND HIS ERUDITION THAN THE SPIRITUAL NEEDS OF SOULS UNTIL THE 1576 PLAGUE IN MILAN CALLED HIM TO HANDS-ON SERVICE. THIS PATRICIAN WADED INTO THE FRAY, MINISTERING TO THE SICK AND ENCOURAGING OTHER PRIESTS TO DO SO AS WELL. FOR WHERE ALL THE WORLD SAW DEATH AND DESO-

---

[136] Roy H. Schoeman, "Catholicism and Judaism," in The Catholic Church and the World Religions: A Theological and Phenomenological Account, ed. Gavin D'Costa (New York, NY: T&T Clark International, 2011), 59.

[137] The official position of the Catholic Church toward the Jews, condemning anit-Semnitism, is articulated in Nostra Aetate, Vatican II, 1965

[138] Elizabeth Lev and Thomas D. Williams, "What the Catholic Church knows about charity in the time of pandemic," angelusnews.com, Mar. 17, 2020, https://angelusnews.com/faith/conversion-in-a-time-of-coronavirus/

LATION, HE SAW A GLIMMER OF POSSIBILITY TO SAVE SOULS."[138]

It was during a subsequent outbreak of the plague that Borromeo, the archbishop of Milan, embraced the call to radical Christian charity, devoted himself to caring for both the material and spiritual needs of the people. At the height of the crisis he was involved in everything from organizing hospitals to feeding thousands of people on a daily basis, and was even responsible for erecting altars on street corners so that Masses could be said where people would be able take part from their balconies. He urged priests to go door-to-door to hear confessions and otherwise tried to make it possible for people to receive the sacraments while remaining quarantined in their homes.[139] Meanwhile, about 50 years earlier in another part of Italy, the Capuchin Order was born from a desire to return to the strict ideal of Franciscan poverty (the Capuchins are an offshoot of the Franciscans), and therefore embraced the opportunities brought by the plague to serve in hospitals and the *lazzaretti*, places where people who had contracted the plague were essentially abandoned to die.[140] These being but a few examples, the list could go on and on. Much like the earliest centuries of Christianity, there have been times and places when people have embraced the call to radical self-giving love, and they have indeed changed the world for the better. Unfortunately, however, what often gets remembered is the times when people have done the exact opposite.

[139] Bullivant, Catholicism in the time of Coronavirus, 30-32.

[140] John Allen Jr., "Capuchin history a reminder that death often brings new Catholic life," Crux Taking the Catholic Pulse, cruxnow.com, Mar. 23, 2020, https://crux now.com/news-analysis/2020/03/capuchin-history-a-reminder-that-death-often-brings-new-catholic-life/

VII.
SPANISH FLU

> "THE FIRST WORLD WAR MAY NOT HAVE CAUSED THE VIRUS IN THE WAY SOME PEOPLE FEARED AT THE TIME. IT WASN'T A MAN-MADE CHEMICAL WEAPON. BUT THE WAR DID GIVE THE VIRUS AN ALMOST ENDLESS SUPPLY OF BODIES, LIVING IN CLOSE QUARTERS, MANY MALNOURISHED WITH ALREADY COMPROMISED IMMUNE SYSTEMS, MANY JUST EMOTIONALLY EXHAUSTED, AND PLENTY WHO HAD NEVER EXPERIENCED A VIRUS LIKE THIS BEFORE. THE RESULT WAS A PANDEMIC LIKE NO OTHER RECORDED, A PANDEMIC THAT KILLED MORE PEOPLE IN SIX MONTHS THAN THE BLACK DEATH DID IN FOUR YEARS. A PANDEMIC WHICH DWARFED THE SACRIFICES OF LIVES DURING BOTH WORLD WARS OF THE TWENTIETH CENTURY AND SHAPED THE CIVILIZATION WE KNOW TODAY."[141]

[141] Jaime Breitnauer, The spanish Flu Epidemic and its Influence on History: Stories from the 1918-1920 Global Flu Pandemic (Great Britain: Pen and Sword History, 2019), 22-23.

Although it was only a century ago, there is much we still do not know about the Spanish Flu epidemic that swept across the world. While we did learn more about it in the decades after the war – as the medical profession confirmed that it was in fact a virus and we steadily discovered more and more about the nature of influenza – and about how it spread in those early moments of the pandemic, the greatest mystery we have yet to uncover, even to this day, are the true origins of the virus. Despite this grey area in the history of the pandemic, we know the impact it had on the world and on the war, as it spread throughout claiming lives and debilitating countries through the number who fell ill during the first wave,[142] and through the number cut down during the second. For some countries during the first World War, the Spanish Flu caused unprecedented levels of change as countries nearly collapsed upon themselves. In others, it drastically changed their military capacity to the point where the Spanish Flu itself could receive partial credit for the allies winning the war,[143] not to mention the effect it had on the peace talks after the first world war – and yet, despite these drastic and costly impacts in countries throughout the world, for many it was almost never spoken of again, as people chose to reflect on the war and the changes it brought rather than think about the far more immediate and personal experiences many in the world had with the influenza pandemic. The aims of this chapter are to discuss the history and origins of the Spanish Flu pandemic, explore the populations affected and the symptoms experienced by those hit by the different waves of the pandemic, and to examine the lasting impact the Spanish Flu pandemic and our later attempts to understand and explain the effect it had on our world today.

The word influenza can be dated back to 16th century Italy, when it was used to describe an epidemic that was believed to have been 'influenced' by an alignment of stars.[144] Influenza pandemics have been occurring regularly throughout our history roughly every 30-40 years since the 16th century,[145] although due to the variable and inconsistent way symptoms and causes of diseases were explained it is difficult to pin down exactly how many of medieval Europe's pandemics were influenza related. Our understanding of influenza has a come a long way since the 1918 Influenza pandemic, where individuals still attempted

[142] Ibid., 28-29.
[143] Ibid., 51.
[144] Ibid., 4-5.
[145] "Lessons Learned from the 1918–1919 Influenza Pandemic in Minneapolis and St. Paul, Minnesota," National Center for Biotechnology Information, Nov.-Dec., 2007, https://www.ncbi.nlm.nih.gov/pmc/articles/PMC1997248/.

to explain influenza as a result of Pfeiffer's bacillus – a type of bacteria isolated by Richard Pfeiffer – despite the fact it wasn't always found in the sick and was sometimes found in healthy individuals. Since then, we have learned that there are four types of influenza viruses, with type A and B viruses causing seasonal epidemics, and type A viruses being the only ones responsible for pandemics.[146] The type A viruses are further broken down into subtypes based on two different surface proteins; hemagglutinin – the H in the shorthand names of the viruses – and neuraminidase – the N in the names.[147] There are 18 types of H (H1 through H18) and 11 types of N (N1 through N11), leading to a potential 198 different influenza A subtype combinations, although only 131 have been seen.[148] Beyond this, the only other details about influenza viruses and how they differ that we need to know for our purposes in this chapter are the topics of antigenic drift and reassortment. Antigenic drift refers to small changes in the antigenic properties of a virus, where antigenic properties concern our bodies ability to recognize a virus and trigger the appropriate immune response[149] – so, although still the same virus, if antigenically different our body will have a harder time fighting it. Reassortment is the process where, when a cell is infected with multiple viruses of the same type, it produces not only the original two strains of the virus, but also a new hybrid strain of the virus called a reassortant[150] a recent example of this being the 2009 H1N1 strain, which was a reassortant of avian, swine, and human influenza viruses.[151] In regards to influenza type A viruses like the Spanish Flu, the process of reassortment leads to the possibility of up to 256 new strains.[152] As we know now, all of this combined is what makes viruses so difficult for our body to fight, since a strain that a person had once had and survived could change to the point where eventually that person's body would have to start all over again in developing an immune response to it, putting them at risk again. However, while

[146] "Types of Influenza Virus," Centers for Disease Control and Prevention, Nov. 18, 2019, https://www.cdc.gov/flu/about/viruses/types.htm.

[147] Ibid.

[148] Ibid.

[149] "Antigenic Characterization," Center for Disease Control and Prevention, Oct. 15, 2019, https://www.cdc.gov/flu/about/professionals/antigenic.htm.

[150] "Reassortment of the Influenza Virus Genome," Virology Blog, Jun. 29, 2009, https://www.virology.ws/2009/06/29/reassortment-of-the-influenza-virus-genome/.

[151] Ibid.

[152] Breitnauer, The Spanish Flu Epidemic and its Influence on History, 17.

we have a better understanding of the inner workings of viruses now, and how to aid our bodies in combating them through the production of vaccines, none of this information was available to the population of the world back in 1918. In fact, it would be another 15 years before anyone confirmed that influenza was a virus in 1933 and was not in fact caused by Pfeiffer's bacillus.[153] This drastic lack of information and understanding, the amount of time it took for the places that did react, and the generally poor conditions for people in many countries due to the war all led to the perfect breeding ground for the 1918 influenza pandemic.

While we know of many early cases and much of how the pandemic spread throughout the world and across the trenches, something we still have yet to determine definitively is the origin of the Spanish Flu. The earliest recorded case that we have available to us is the case of Albert Gitchell, a mess cook who fell ill March 1918 while serving at Camp Funston, located on Fort Riley in Kansas. Although considered patient zero by some, all we can know with certainty about Gitchell's case is that he was simply the first recorded case that we know was the Spanish Flu,[154] though, this hasn't stopped some from positing Kansas as the origin of the pandemic.[155] Another commonly posited origin for the pandemic is the northern Chinese province of Shanxi. During the winter of 1917, Dr Wu Lien-teh – a celebrated Chinese physician who had previously addressed the plague outbreak of 1910 in China and who laid the foundation for a public health system in China[156] – received word of a disease with similar symptoms to the plague he had dealt with nearly a decade prior. However, unlike the previous outbreak of plague, there were not nearly so many people dead, and it was this fact, along with a lack of corroborating evidence and political motivations, that led to the Chinese government dismissing Dr Wu's insistence that this Winter Sickness was another case of plague.[157]

[153] Ibid., 16.
[154] Ibid., 4.
[156] SciShow, "The 1918 Pandemic: The Deadliest Flu in History," Youtube Video, 6:08, Jan. 7, 2018, https://www.youtube.com/watch?v=u7xIGcLGTu8.
[157] Breitnauer, The Spanish Flu Epidemic and its Influence on History, 6-8.

By the spring of 1918, the winter sickness had passed, and was soon forgotten in the wake of all the wartime happenings. A third proposed origin, one intimately connected to the former, is 1916 France, where at the time physicians were dealing with a disease that they would later say was "fundamentally the same condition" as the Spanish flu epidemic,[158] many of them expressing the sense of déjà vu they experienced fighting against the pandemic and how similar it was to what they were dealing with in Northern France from 1916 to 1917.[159] Yet another related origin is the British military camp in Étaples. At the time there was a bronchial pneumonia going through the trenches, and many people at the camp were ill.[160] Before long, soldiers were dying of what again had a resemblance to the later confirmed cases of the influenza pandemic. What connects these last three potential origins though, beyond the obvious connection of the British base being located in France at a time where there was a Spanish Flu like virus, is the Chinese Labour Corp. From 1917-1918, China sent nearly 100,000 men to France and England, mainly by shipping them to Canada where they were then taken by railcar to the east coast before finally being shipped off to Europe to support the war effort,[161] with more having been sent since 1916 to France and Russia.[162] While far from conclusively determining China as the origin for the disease – since, not only could the spread actually have been in the opposite direction, with returning people from the labour corp bringing the pandemic home opposed to spreading it originally, but it also does little to explain the early cases in Kansas amongst soldiers who had yet to deploy – the early and far movements of the Chinese labour corp connects these potential origins, and, at the very least, provides possible avenues the virus could have spread once it began –

---

[157] Ibid., 8-9.

[158] "The Influenza Pandemic of 1918," University of Oxford, http://wwlcentenary.oucs.ox.ac.uk/body-and-mind/the-influenza-pandemic-of-1918/.

[159] Ibid.

[160] Breitnauer, The Spanish Flu Epidemic and its Influence on History, 10-11.

[161] "1918 Flu Pandemic That Killed 50 Million Originated in China, Historians Say," National Geographic, Jan. 24, 2014, https://www.nationalgeographic.com/news/2014/1/140123-spanish-flu-1918-china-origins-pandemic-science-health/.

[162] Breitnauer, The Spanish Flu Epidemic and its Influence on History, 10-11.

[163] Ibid., 4.

[164] Ibid., 99.

[165] "1918 Flu Pandemic That Killed 50 Million Originated in China, Historians Say," National Geographic, Jan. 24, 2014, https://www.nationalgeographic.com/news/2014/1/140123-spanish-flu-1918-china-origins-pandemic-science-health/.

not that the virus needed any more help spreading than it already had with the constant wartime movements of soldiers from the front line to home and back again. This uncertainty about the origins of the pandemic have led to a lot of finger pointing, both during the pandemic and into the present. Early origin theories from 1918 had Americans laying blame on China or France[163] – even though, if anything, America infected France themselves – and China saying it had been infected from workers coming back from Russian poppy fields.[164] To this day some still blame China as being the start of the 1918 pandemic,[165] while some others now put the blame on America, where some of the earliest cases were noticed,[166] and others France.[167] Despite all of these possibilities though, we don't know where the pandemic originated, and it is entirely possible that we never will. However, while definitely a mystery worth solving if possible, our lack of knowledge concerning its origins does nothing to take away from the impact that the pandemic had, or the number of people affected, and lives lost.

The Spanish Flu pandemic was the worst pandemic the world had seen, and estimates on how many people died differ greatly. More conservative estimates put the number of deaths to be around the 25 million mark,[168] though others estimate the number to be more than 50 million,[169] and others still estimate the death toll to be potentially as high as 100 million.[170] However, these deaths weren't uniformly spread over time, or over the world, as the pandemic came in three waves: the first wave began early March of 1918 – although as previously discussed the origin and patient zero are unknown, and there are potential cases of it as early back as 1916 – and quickly spread throughout Europe, reaching as far as Poland by July.[171] The second and by far deadliest wave[172] began in the summer and lasted until just before Christmas, with many of those falling ill dying within two days of the first

[166] "Lessons Learned from the 1918–1919 Influenza Pandemic in Minneapolis and St. Paul, Minnesota," National Center for Biotechnology Information, Nov.-Dec., 2007, https://www.ncbi.nlm.nih.gov/pmc/articles/PMC1997248/.

[167] "The Influenza Pandemic of 1918," University of Oxford, http://ww1centenary.oucs.ox.ac.uk/body-and-mind/the-influenza-pandemic-of-1918/.

"Influenza pandemic of 1918-19," Encyclopedia Britannica, Jul. 7, 2020, https://www.britannica.com/event/influenza-pandemic-of-1918-1919.

[168] "Lessons Learned from the 1918–1919 Influenza Pandemic in Minneapolis and St. Paul, Minnesota," National Center for Biotechnology Information, Nov.-Dec., 2007, https://www.ncbi.nlm.nih.gov/pmc/articles/PMC1997248/.

https://www.youtube.com/watch?v=u7xIGcLGTu8

[169] "Influenza pandemic of 1918-19," Encyclopedia Britannica, Jul. 7, 2020, https://www.britannica.com/event/influenza-pandemic-of-1918-1919.

[170] Ibid.

[171] "Influenza pandemic of 1918-19," Encyclopedia Britannica, Jul. 7, 2020, https://www.britannica.com/event/influenza-pandemic-of-1918-1919.

symptoms[173] – in fact, the second wave hit so much harder than the first wave or the normal seasonal flu that rumours spread that it was a form of chemical warfare, rumours which are understandable given how both the chemical weapons in use and the flu itself damaged the lungs of victims. It's even been proposed that chemical weapons could have been the cause of the violent change in the virus between the first and second waves, since the weapons could cause genes to mutate, though this theory couldn't ever be confirmed over a century down the line.[174] Then, there was a moment to breathe, but only a short one, for by January, with the demobilization of troops bringing everyone home a third wave began,[175] with this and the second wave claiming the majority of the pandemic's victims. Additionally, countries that had suffered more from the first wave and that had had wider spread infection tended to have lower mortality rates during the later waves, presumably because of a modest amount of developed immunity to the virus.[176] In spite of all the death just from the flu alone leading up to the armistice, when it was declared thousands gathered to celebrate, causing the death toll to spike shortly after.[177] With a virus this deadly, one would think that it would have gotten a name befitting its impact on the world. However, instead of being given a moniker that would fit alongside the black death which came before it, the influenza pandemic of 1918 was called the Spanish Flu, or the Naples soldier in some places, and neither for particularly serious reasons. Its most well known name – the Spanish Flu – came from the wartime censorship present in the warring countries of the world at the time. Although Spain was anything but the first to be affected by the flu, they were the first to openly report on it since, being neutral, they weren't limited to only reporting more positive things to their people, thus leading the rest of the world to dub the virus the Spanish Flu.[178] Its other common moniker, one limited primarily to Spain itself, came from

[172] Ibid.
[173] Ibid.
[174] Breitnauer, The Spanish Flu Epidemic and its Influence on History, 21.
[175] Ibid.
[176] Ibid., 99-100.
[177] Ibid., 31.

a comparison being made between the virus and a particularly catchy, well known song from the time. Because of how widespread and catchy the tune 'The Soldier of Naples' – from the operetta The Song of Forgetting – was, one newspaper described the virus as being just as catchy, leading the people of Spain to refer to it as The Naples Soldier.[179] Despite the names though, this version of influenza was no seasonal flu, and was certainly not something to be taken lightly. The Spanish flu ran rampant through the world, tearing families apart and killing so many that higher estimates of the toll put the number above all the lives lost from both world wars put together, and which claimed more lives in those few months of the second wave than the black death claimed over years.[180] The Spanish Flu was a monster, and it left a wake of carnage and bodies wherever it went.

The Spanish Flu spread to nearly every inhabited part of the world as it ran its course, and while it certainly didn't play favorites regarding who it infected, it wasn't exactly indiscriminate concerning who it killed. Military personnel were some of the first to become infected and to fall victim to the flu,[181] and movements of military personnel both on and off the battlefield were one of the main ways the flu traveled. During the first wave, the casualties were primarily from the usual age groups for a disease: the very young and the very old, resulting in a typical U shape when graphing the age groups at greatest risk.[182] However, the later waves of the pandemic had something unique to them beyond their increase in mortality rates: the second and third wave killed about as many people from the ages of 20-40, the age range normally considered the healthiest and at lowest risk, as from the other two highest risk groups. In fact, about as many as half of all the deaths from the later two waves were from this normally particularly healthy age group.[183] This strange trend was determined to actually be caused by how healthy the younger victims were, as the

[178] Ibid., 29.
[179] Ibid., 41.
[180] Ibid., 23.
[181] "1918 Pandemic (H1N1 Virus)," Centers for Disease Control and Prevention, Mar. 20, 2019, https://www.cdc.gov/flu/pandemic-resources/1918-pandemic-h1n1.html
[182] Breitnauer, The Spanish Flu Epidemic and its Influence on History, 17.

body's immune response would be so strong that it would lead to something called 'cytokine storms.' Essentially, the immediate and powerful immune response from the body would lead to too many immune cells flooding an infection site and causing inflammation; when this site of infection was the lungs – which it often was since this is where the flu itself did most of its damage – then the patient's lungs would fill due to the inflammation, causing people to often "drown in their own bodily fluids."[184] In other words, the virus "turned the body of the healthiest of victims on itself."[185] In addition to normally healthy, the flu also disproportionately affected pregnant women, as well as women in general; this, combined with wartime casualties and young men killed by the flu led to whole families being torn apart, and a sharp increase in the number of orphans, as well as to 1918 being the first year recorded where there were more deaths than births.[186] To say that the Spanish Flu had an impact on the world would be an understatement. Beyond the brutal and often graphic ways it tore families apart, driving people to make attempts at their families and their own lives out of despair,[187] when not simply killing the majority of a family itself.[188] It shut down countries on both sides of the war as it went, with the first wave crippling them through the number sick before the second wave swept through and killed, impairing military movements and potentially leading Germany to lose the war, according to some.[189] The heavier death toll amongst Black communities compared to those of the White Christian colonists led to many communities turning towards religion. However, disappointed with the response from European churches, Black Christian communities were founded instead across South Africa, with around 85% of people in South Africa identifying themselves as Christian today.[190] In India, though Gandhi himself fell ill for a time, supporters of his movement towards equality and independence, and organizations that had aided him worked to help the people of

[183] "Influenza pandemic of 1918-19," Encyclopedia Britannica, Jul. 7, 2020, https://www.britannica.com/event/influenza-pandemic-of-1918-1919.

[184] Breitnauer, The Spanish Flu Epidemic and its Influence on History, 20.

[185] Ibid., 21.

[186] Ibid., 32-33.

[187] Ibid., 111-112.

[188] Ibid., 31-32, 39-40, 53.

[189] Ibid., 28-29.

[190] Ibid., 62.

India in a time where they felt the government had failed to act, and abandoned them, ultimately strengthening the drive for independence as the only significant response from the British government was not to aid the people against the flu, but instead to extend wartime legal restrictions, suspending civil liberties past the end of the war.[191] In the north, First Nations communities suffered greatly, losing many leaders, hunters, and families in their entireties, to the point where recovery and return to their traditional lifestyle seemed impossible. Worse still, after the flu had run its course, some communities received the ultimatum from officials that widows and widowers were to couple in order to take care of the many orphans, or the Inupiat children would be sent south and be institutionalized, leading to a devastation in communities not only in numbers, but in belief.[192] Although the Spanish Flu ended up claiming more lives than the war, even when looking at the smallest estimates of casualties from the pandemic, much of the world was particularly slow to react to it. For many countries, the war took precedence. In England, despite the efforts of a mathematician named Major Greenwood to warn officials of the impending second wave, all he received for his efforts was derision, as individuals such as the General Medical Council representative and Army Sanitary Committee representative Sir Arthur Newsholme dismissed his concerns, as he felt it was more important to keep the country functioning and to prioritize the war effort[193] – an ironic sentiment in hindsight, given how the flu brought England to its knees for a time. Over in the US, although some cities, such as Washington DC and San Francisco, responded quickly to the pandemic, others like NYC were again driven more by concerns over keeping the city moving and with the war effort, and as such failed to respond until October, ultimately with responses still less restrictive than in other cities.[194] What's worse, barely a week after the city finally took some form of action,

[191] Ibid., 84-86.
[192] Ibid., 66-68.
[193] Ibid., 29-30.

discouraging large gatherings and other safety risks, President Woodrow Wilson was allow to take a procession of roughly 25,000 people through NYC; 2000 people died in the city that same week, which is more deaths than cities like San Francisco suffered from the flu over the entirety of the pandemic.[195] This response somehow pales in comparison to the reaction of Philadelphia officials, who were practically in a state of denial about the flu, with the inaction there leading to almost as many deaths as NYC, despite having less than half the population.[196] However, for every case where a city refused to act until the damage had already been done, or a population refused to follow guidelines despite quick reaction from their governments, ultimately resulting in unnecessary deaths, there were individuals doing all they could to help. In India, as previously mentioned, Gandhi's supporters stepped in when the government failed or effused to do so themselves, saving lives and risking their own in order to help. Efforts like these could be found across the world, in many places with similar levels of near criminal inaction. In Philadelphia, despite the awful incompetency of the official responses, over 2000 nuns were on the street attempting to help, tending to the sick and groups less likely to be able to see a doctor.[197] Likewise in England, Rose Selfridge and her husband Harry Selfridge turned their home at Highcliffe Castle in Dorset into a hospital where US soldiers could be treated because of Rose's insistence that they could be doing more with their wealth to help, and with Rose personally working there despite her own health concerns.[198] Rose Selfridge lived to help, putting on recitals to raise money for charities, and working at the Christchurch hospital before helping at the one they made. In the end, Rose Selfridge died helping people as best she could. Being prone to respiratory illnesses, her close contact with soldiers suffering from the flu meant it was only a matter of time before she would catch it too, and at the beginning of May she

[194] Ibid., 58-59.
[195] Ibid., 59.
[196] Ibid., 61-62.
[197] Ibid., 62.
[198] Ibid., 26-27.

caught the Spanish Flu, dying of pneumonia on the 12th.[199] In the face of inaction from those obligated to help, people like Rose stepped up, risking and, in Rose's case, giving their lives to help as best they could to fight back the pandemic. In times of hardship and tragedy, it's easy to forget that people like Rose are out there, and to get lost in the despair of the situation. Going forward from the wartime world that suffered the brunt of the first and second waves of the pandemic, one can only imagine how different things would have turned out if more people gave to such an extent to help others.

After the war had ended, the effect of the flu continued to snowball; though the worst of the flu was over, some delegates at the peace conference still fell ill, including Woodrow Wilson. While these bouts of the flu likely contributed to some of the decisions made at the conference, Wilson's illness contributed to a stroke later on, leading to him removing himself from public life, and preventing him from persuading the US government to join the League of Nations or consent to the Treaty of Versailles. Worse yet, although Wilson supported laying the blame for the war on Germany, he hadn't supported the punitive measures that would eventually be put upon Germany, and which would eventually snowball to "the backdrop of humiliation and desperation"[200] used by Adolf Hitler in order to rise to power. In this way, the Spanish Flu bears some of the blame for the situation in Germany leading up to the second world war. The flu also set the groundwork for true public health systems in the wake of the many failures of those in place during the flu in places such as England, France, and Russia.[201] In Brazil, the flu led to the Sanitation Movement[202] and furthered Dr Wu in establishing a public health system in China.[203] The Spanish Flu led to increases in depression, anxiety, and feelings of helplessness, and left survivors of it with numerous potential aftereffects, many of which

[199] Ibid.
[200] Ibid., 110-111.
[201] Ibid., 35-36.
[202] Ibid., 65.
[203] Ibid., 100-101.

wouldn't be realized for years yet to come. For soldiers who survived both the war and the flu but suffered from PTSD, it was noted that they were more likely to suffer from mental impairment and a number of neurologically based issues. As the generation affected by the Spanish flu aged and medicine advanced, further aftereffects were able to be detected. Studies showed a range of effects, including a spike in cases of cardiovascular disease, increased rates of heart disease in those affected by the flu in utero, and difficulties in school and in holding down jobs for those exposed to the second wave prenatally.[204] Despite all the pain and devastation it caused – or, maybe for that exact reason – one of the strangest things about the legacy of the Spanish Flu is the lack of one. Theories as to why vary; some suggest that it was forgotten in a conscious attempt to dull the pain it caused; others suggest it was overshadowed by the war for one reason or another; and others still argue that, perhaps it wasn't any sort of intentional purge or silence at all, but was instead simply the natural reaction from a world that had already suffered more than it could bear.[205] Regardless as to why the world was so silent about the flu in the years following it, the influenza pandemic of 1918 is not one that will be truly forgotten anytime soon, as it left its mark on the world in the form of 100 million graves.

[204] Ibid., 112-114.
[205] Ibid., 114-115.

VIII.
SPANISH FLU AND CHRISTIANITY

> *Pandemics are scary. Much of what terrifies us is the lack of control that disease and its attendant consequences (quarantine, disruption, uncertainty) bring. The medieval world knew it had little control over much of anything — nature, weather, war, overlords — but the contemporary world is accustomed to a feeling of self-determination. A society used to thinking it controls everything from climate to gender is poorly equipped to weather the storm of unpredictability that epidemics bring.*"[206]

[206] Elizabeth Lev and Thomas D. Williams, "What the Catholic Church knows about charity in the time of pandemic," angelusnews.com, March 17, 2020, https://angelusnews.com/faith/conversion-in-a-time-of-coronavirus/

Much like the Black Death, the consequences of the Spanish Flu were absolutely catastrophic. Unlike the Black Death, however, the Spanish Flu pandemic of 1918 not only coincided with the First World War but also took place around 200 years after the so-called 'Enlightenment', resulting in a very different response on the part of both the Church and society as a whole; which, when compared to the time of the Black Death, had become increasingly distinct from one other. Some of the same questions were being asked, and some of the same explanations were being proposed in the early 20th century as they were in the 14th, however, with the birth of the modern sciences there was a definite shift towards what we might consider a more 'realistic' approach to natural disasters. This doesn't mean that theological questions, such as why would God permit evil in the first place, were no longer being considered, but it does mean that even people of faith began to recognize that there may have been more to the answer than meets the eye. With that in mind, in this chapter we'll be taking a quick look at how the society affected by the Spanish Flu was very different from the one affected by the Black Death; in addition to delving a little deeper into the question of how to reconcile the notion of a God who is love with the realities of suffering and evil, or more precisely, how did this affect the way Christians responded in the wake of the Spanish Flu. Were they still able to see the hand of God at work in the midst of such a terrible catastrophe?

In addition to coinciding with the First World War, the Spanish Flu also roughly corresponded with what some consider to be the transition from modernity to postmodernity. Following years of cultural upheaval and all sorts of changes brought about by the so-called Enlightenment, even though Western society in general still acknowledged its Christian roots, it was far from the predominantly Christian society of the high middle-ages. As medicine, science and technology continued to make significant advances, people were becoming increasingly likely to look to scientific explanations, while at the same time, ready to discount religious ones. Much of what might now be considered "modern" thought had been thoroughly shaped by the concept of a 'freedom of indifference' stemming from Ockham's nominalism, having essentially sown the seeds of moral relativism; not necessarily bearing the kind of fruit as can be seen today, but nonetheless very much present. The sort of religious skepticism and empirical rationalism stemming from the Enlightenment was laying the groundwork of what today might be defined as 'scientism', the reduction of all forms of knowledge to the scientific form of knowledge; which, of course, leaves less and less room for the transcendent. Therefore, when it came to answering age-old questions such as the problem of evil, the traditional theological answers were not given nearly as much weight as they had in the past. The predominant theological trend at the time was a 'neo-scholasticism' that, having embraced the great precision with which the scholastics had approached both faith and reason and reason, unfortunately tended toward a somewhat limited or narrow view of both theology and the sciences. Even though it may have had its merits, neo-scholasticism was incapable of dialoguing with an increasingly secular culture and resulted in religious explanations being that much easier to discount. As we'll see in a subsequent chapter, in the years following World War I and the Spanish Flu a new approach to theology was being developed that was better prepared to meet the 'modern' world, more-or-less on its own terms, and find new ways to share the message of the gospel in-light of scientific progress. The world afflicted by influenza in 1918 was going through a significant turmoil on many different levels. The Christian foundations of western civilization were not necessarily crumbling, but definitely being called into question. So how, it might be asked, did people respond?

As terrible as the Spanish Flu was, it did bring about more than just suffering and death. Reflecting on some of the effects of the 1918 pandemic, author Jamie Breintnauer puts it this way:

"WE SAW DESPERATION, CONFUSION AND GRIEF, BUT WE ALSO SAW TENDERNESS AND A MOVE TOWARD SOCIAL CARE. WE SAW PEOPLE REACHING FOR EACH OTHER ACROSS BARRIERS OF CLASS, CULTURE AND LANGUAGE TO OFFER A HELPING HAND. WE SAW THE OPPRESSED SUPPORTING THEIR FLAILING OPPRESSORS IN THE QUEST FOR SURVIVAL AND WE SAW THOSE IN POWER MAKING DECISIONS TO BENEFIT THE MANY, NOT THE FEW IN A

> TIME WHEN THE WELFARE STATE WAS UNHEARD OF. ONE OF THE KEY FEATURES OF THE PANDEMIC WAS THE POWER OF COLLECTIVE MOVEMENTS AND UNDERSTANDING THAT, AS A UNITED GROUP, THIS TERROR WAS SOMETHING THAT COULD BE CONQUERED. WHILE SCIENCE WORKS TO TRY AND PREVENT SUCH A DISASTER FROM REPEATING, SOCIETY CONTINUES TO EVOLVE DOWN THE PATH THAT WAS SET BY SPANISH FLU, A PATH WHERE ACCESS TO HEALTHCARE, SAFE HOUSING AND CLEAN WATER HAVE BEEN RECOGNIZED AS A HUMAN RIGHT, A PATH WHERE COLLABORATION IS CELEBRATED AND A PATH WHERE GROUP COOPERATION FOR MUTUALLY BENEFICIAL OUTCOMES ARE SEEN AS THE NORM."[207]

It is interesting, as Breintnauer points out, that widespread devastation is actually capable of bringing out both the best and worst in people. In many ways the dual catastrophe of a world-wide pandemic combined with the devastation left by such a large scale war left many people wondering what was the appropriate response? During the Black Death many people's religious convictions led them to believe that God was punishing people for their sin, and this sort of mentality certainly wasn't absent during the Spanish Flu either:

> "INDIVIDUALS OF VARIOUS RELIGIOUS AFFILIATIONS ATTRIBUTED THE PANDEMIC TO GOD'S DIRECT ACTION, EITHER AS A PUNISHMENT OR FOR SOME DIVINE PURPOSE. SOME FELT THAT THE SINS BEING PUNISHED BY INFLUENZA WERE HIGHLY SPECIFIC, INCLUDING A STRAYING FROM GOD, INDIFFERENCE, UNBELIEF AND SUPERSTITION, FAILING TO ATTEND CHURCH, AND THE WORSHIPPING OF SCIENCE. OTHERS FELT THAT THE PANDEMIC WAS A DIVINE MEANS FOR ENDING A DEVASTATING WAR."[208]

Looking back, however, we might recognize how it was that this particular period contributed to the greater

[207] Jamie Breitnauer, The Spanish Flu Epidemic and its Influence on History: Stories from the 1918-1920 global flu pandemic (Philadelphia, PA: Pen and Sword Books Ltd, 2019), 122.

[208] Virginia Aronson, The Influenza Pandemic of 1918 (Philadelphia, PA: Chelsea House Publishers, 2000), 75.

good in the way that it reflects the gradual process of growth that human beings have been undergoing since the very beginning. In many ways, what might be described as 'growing pains'—or from a theological standpoint, the concept that "grace perfects nature."[209] Even if many people may have been losing something of their 'religiosity', they might have gained something of their 'humanity'. As we turn to consider the question of why God permits evil, we'll see how this played out in the wake of the Spanish Flu with a couple of concrete examples.

If God really is infinite goodness, why does evil exist? In the words of St. Thomas Aquinas, citing St. Augustine before him:

> "AS AUGUSTINE SAYS: SINCE GOD IS THE HIGHEST GOOD, HE WOULD NOT ALLOW ANY EVIL TO EXIST IN HIS WORKS, UNLESS HIS OMNIPOTENCE AND GOODNESS WERE SUCH AS TO BRING GOOD EVEN OUT OF EVIL. THIS IS PART OF THE INFINITE GOODNESS OF GOD, THAT HE SHOULD ALLOW EVIL TO EXIST, AND OUT OF IT PRODUCE GOOD."[210]

As simple as this explanation might sound on paper, anyone with firsthand experience of evil will be the first to tell you, it doesn't offer a whole lot of consolation in the heat of the moment. It is, however, an integral part of the Christian mystery. For at the very heart of the Christian faith is the Paschal mystery— the passion, death and resurrection of Jesus. Why would God permit his own Son to experience the full effects of sin and death when he had done absolutely nothing to deserve it? Again, the simple answer: because he was going to bring about a greater good—in this case, our redemption. Therefore, from the very beginning Christians have understood that suffering can be embraced as a share in Christ's redemptive suffering: "I am now rejoicing in my sufferings for you sake, and in my flesh I am completing what is lacking in Christ's afflictions for the sake of his body, that is, the church." (Col 1:24) Up and down the centu-

---

[209] St. Thomas Aquinas, Summa Theologiae, I Q. I, A. 8, ad 2.
[210] Ibid., I Q.2, A.3, ad I.

ries there have been countless examples of people who have willingly embraced horrific suffering as a share in the cross of Christ, and the opportunity presented by the Spanish Flu was no exception. For example, in 1917 three young shepherd children reportedly received a series of miraculous apparitions of the Blessed Mother in Fatima, Portugal.[211] Not long afterwards, Francesco and Jacinta, two of the visionaries (since been canonized by the Catholic Church), both died of influenza. Having been told by Our Lady they were soon to die they "gave themselves more and more to mortification and prayer", offering their suffering for "the love of God, the conversion of sinners and in reparation for the offences committed against the Immaculate Heart of Mary."[212] The idea that suffering could be offered up on behalf of another stemming from the radical gift of self that Jesus made on the cross: "No one has greater love than this, to lay down one's life for one's friends." (Jn 15:13) Even though they were still only children, part of what these two young saints understood is how valuable suffering can be when it is united to Christ's. The question is not if we are going to suffer, the question is what we do when we suffer. Dr. Peter Kreeft describes it as follows: "What then is suffering to the Christian? It is Christ's invitation to us to follow him. Christ goes to the cross, and we are invited to follow… Suffering in not the context that explains the cross; the cross is the context that explains suffering."[213] As difficult as it might be to grasp the concept that an all good, all loving God would will, or even permit, that anyone suffer; again, it must be recognized in the context of the Christian mystery: "God can bring great good out of the consequences of an evil. He used the Spanish flu to shape two little saints and to let them participate in the work of saving sinners."[214] Another authentic 'Christian' response to suffering can be seen in the witness of numerous religious women who served the sick during the Spanish Flu. In the United States in particular there are accounts of "thousands of women

[211] Even though the Catholic Church officially approves of the Fatima Apparitions, it does not consider it obligatory that all the faithful acknowledge this event as part of the deposit of faith. See the Catechism of the Catholic Church no. 67

[212] John de Marchi, The True Story of Fatima: A complete account of the Fatima Apparitions (Fort Erie, ON: The Fatima Center, 1947), 60-61.

[213] Peter Kreeft, Making Sense out of Suffering (Ann Arbor, MI: Servant Books, 1986), 137.

[214] Anthony Ho, "How God brought good from a pandemic," The B.C. Catholic, March 18, 2020, https://bccatholic.ca/voices/father-anthony-ho/how-god-brought-good-from-a-pandemic

religious (who) took on nursing duties in hospitals or clinics and went into private homes to offer food, medicine, comfort and even housecleaning to families affected by the Spanish flu."[215] homes to offer food, medicine, comfort and even housecleaning to families affected by the Spanish flu." In the words of Boston Health Commissioner William C. Woodward:

> "WHILE PRAISE MUST BE GIVEN TO ALL WORKERS WHO ASSISTED IN FIGHTING THE EPIDEMIC, SPECIAL PRAISE SHOULD GO OUT TO THE CATHOLIC SISTERS WHO WENT OUT FROM THEIR CONVENTS INTO PRIVATE HOMES AND, WITH THE UNSELFISHNESS THAT CHARACTERIZES THESE WOMEN..., GAVE SERVICE WHICH MONEY ALONE COULD NEVER HAVE PURCHASED. THESE DEVOTED WOMEN, MANY OF THEM TEACHERS AND NOT AT ALL USED TO NURSING, NEVER HESITATED TO PERFORM SERVICES WHICH ARE THE DUTY OF THE PROFESSIONAL TRAINED NURSE... WHATEVER WAS NEEDED TO BE DONE IT WAS THEIR PLEASURE TO DO."[216]

In Philadelphia, for example, one of the hardest-hit cities in the United States, where an estimated 13,000 to 16,000 people died from influenza, nearly 2,200 non-cloistered religious sisters took some role in fighting the epidemic.[217] Tackling the Spanish Flu with the same dedication with which the Christians of the first several centuries slowly transformed the Roman Empire all those years ago, the Loretto Sisters appeared to have had a similar effect amongst mining communities in eastern Kentucky. After her experience working among the miners, many of which harboured strong anti-Catholic sentiments, Sister Mary Gabriel Berry reported: "our going among them on our errand of mercy wrought a change,"and "Many of them said to us, 'We will never again believe such tales and terrible things about Catholics.'"[218] What these religious sisters demon-

---

[215] Nancy Frazier O'Brien, "Sister's work during 1918 Spanish Flu seen as model for crisis today", Vermont Catholic, March 31, 2020, https://vermontcatholic.org/nation/sisters-work-during-1918-spanish-flu-seen-as-model-for-crisis-today/

[216] Ibid. - citing article in Oct. 18th, 1918 edition of the Boston Traveler newspaper

[217] Ibid.

[218] Ibid.

strated is that Christianity does have the potential to have the same sort of influence on modern society as it did in ancient Rome, however, sometimes it takes something drastic, like a pandemic, to really see it.

Was medieval Christendom the 'ideal' society? Not likely. If it were, the normal response to the Black Death would not have been running for one's life but caring for one's neighbour. Just as we saw some examples of this in the wake of the bubonic plague, so too during the influenza outbreak of 1918. Up and down the centuries there have been people who have embraced the call to radical Christian charity—even as it has taken different forms in the face of different types of evil—as well as those who have not. Even though it is often easier to focus on the negative, those who didn't do the right thing, it is well worth considering what the world would be like today without those who did. That being said, the question may not be why would a good God permit evil. The real question just might be, do we have the eyes to recognize the greater good when we see it?

IX.
FAMINES AND CHRISTIANITY

> ❝ I WENT TO ONE VILLAGE AND SAW 100 CORPSES, THEN ANOTHER VILLAGE AND ANOTHER 100 CORPSES. NO ONE PAID ATTENTION TO THEM. PEOPLE SAID THAT DOGS WERE EATING THE BODIES. NOT TRUE, I SAID. THE DOGS HAD LONG AGO BEEN EATEN BY THE PEOPLE.❞
> -YU DEHONG, 1959[219]

[219] Mark O'Neill, "A hunger for the truth," South China Morning Post, July 6, 2008, https://www.scmp.com/article/644236/hunger-truth.

[220] J Colette Berbesque et al., "Hunter-gatherers have less famine than agriculturalists," Biology letters 10, no. 1, (Jan. 8, 2014), doi:10.1098/rsbl.2013.0853

The history of famine is inseparable from the history of structured society. For as long as people have lived in villages, towns, or cities they have experienced famine. Research has suggested that early hunter-gatherer communities probably experienced famine less frequently than later agriculturalist societies, the idea being that while farming populations may enjoy great amounts food in times of good harvest, hunter-gatherer populations are less reliant on favorable growing conditions and thus are more able to adapt to famine and drought.[220] The mental image which typically springs to mind when considering famine is that of crops being ruined by some uncontrollable force of nature, such as flood, blight, drought, or bad weather. This simple view may be adequate for many famines throughout human history, but ultimately, in order to discuss some of the more devastating famines to have occurred in the past millennium, it must be acknowledged that the causes behind famines are often far more complicated than simply "bad weather" or "forces of nature." To that end, the distinction must be made that the term "famine" does not simply refer to the situation wherein people starve due to a shortage of food to eat, but the more broad definition of "widespread starvation caused by a lack of access to food to eat." For example, "famine" can refer to both the scenario where people starve because their nation has literally ran out of food, and the situation where the population of a country starves while the government hoards massive stockpiles of food. With this distinction made, we may now discuss the complex social, political, and/or economic causes behind some of the most devastating famines in recent history.

As we cover specific famines, we will avoid discussing the ground-level reactions and individual experiences of people in each famine. The reason being that the experience of starvation does not sufficiently differ from one location or time period to another to warrant repeated discussion. That being said, it is very important to underline exactly how painful and dehumanizing the experience of dying from starvation is before we move forward. The commonly repeated fact that, "the human body can survive for three days without water and thirty without food," is reasonably accurate, but obscures the extreme lengths which the human body will go to in order to survive without food. These processes undertaken by the body can be roughly divided into three phases. In the first phase, blood glucose levels are maintained by the breakdown of fat and protein stores in the skeletal muscles as well as glycogen in the liver. In the second phase, energy is mainly produced from the breakdown of a person's fat stores. These first two phases can occur even during controlled periods of dieting or fasting, while the third is far and away the most dangerous, and a clear indication that something is very wrong. Once a body's reserves of fat have been depleted, the body turns to the final source of protein in the body: the muscles. The body metabolizes muscle tissue, leading to massive losses in muscle mass. By this point a person will have become weak, apathetic, irritable, and extremely immunocompromised. The way in which starvation directly kills a person is from a heart attack or cardiac arrhythmia. More often than heart failure, however, starvation kills from secondary infections brought on by a compromised immune system. If you take that individual physical reaction to starvation and extend it to an entire population, it becomes easier to understand the kind of complete disintegration of normal society that famine causes. Overwhelming feelings of hunger in the first days of starvation eventually disappear as the body and mind become so weak that a person can't function normally, and eventually a point of no return is reached where the body is so dehydrated and starved of energy that not even food can save a person.[221] We could go on and talk about generational effects of long-term malnutrition as opposed to immediate starvation, but ultimately we simply want to emphasize the fact that the individual effects of any famine are just as cruel and painful as any other infectious disease, if not more so.

The most simple and popular answer to the question of why the Irish Potato Famine occured goes something along the lines of: "The Irish relied too heavily on potatoes for their food and their potato crops were destroyed by blight." This answer is factually correct, but it betrays the complexity behind the famine and, as we will discuss, how foreseeable and possibly preventable the Irish Potato Famine ultimately was. On January 1st, 1801, the Kingdom of Ireland and the Kingdom of Great Britain were officially merged to become the United Kingdom of Great Britain and Ireland. With this merging the Parliament of Ireland was dissolved and the British Government assumed control. During this time, Ireland faced significant economic turmoil, with a significant portion of the population living in abject poverty. In 1843, two years before the beginning of the famine, a Royal Commision was established by the British Government in order to investigate possible land lease issues in Ireland. William Courney, the

---

[221] Thomas C. Weiss, "What Happens When We Starve? Phases of Starvation," Publications, Disabled World, last modified October 29, 2018, https://www.disabled-world.com/fitness/starving.php.

Earl of Devon and the head of the commission, described the state in which the rural population of Ireland lived as follows:

> "THAT THE AGRICULTURAL LABOURERS OF IRELAND SUFFER THE GREATEST PRIVATIONS AND HARDSHIPS; THAT THEY DEPEND UPON PRECARIOUS AND CASUAL EMPLOYMENT FOR SUBSISTENCE; THAT THEY ARE BADLY HOUSED, BADLY FED, BADLY CLOTHED, AND BADLY PAID FOR THEIR LABOUR; THAT IT WOULD BE IMPOSSIBLE TO DESCRIBE ADEQUATELY THE SUFFERINGS AND PRIVATIONS WHICH THE COTTIERS AND LABOURERS AND THEIR FAMILIES IN MOST PARTS OF THE COUNTRY ENDURE; THAT IN MANY DISTRICTS THEIR ONLY FOOD IS THE POTATO, THEIR ONLY BEVERAGE WATER; THAT THEIR CABINS ARE SELDOM PROTECTION AGAINST THE WEATHER; THAT A BED OR BLANKET IS A RARE LUXURY; AND THAT NEARLY IN ALL, THEIR PIGS AND THEIR MANURE HEAP CONSTITUTE THEIR ONLY PROPERTY; THAT A LARGER PROPORTION OF THE ENTIRE POPULATION COMES WITHIN THE DESIGNATION OF AGRICULTURAL LABOURERS, AND ENDURE SUFFERINGS GREATER THAN THE PEOPLE OF ANY OTHER COUNTRY IN EUROPE HAVE TO SUSTAIN."[222]

[222] Daniel O'Connell and Mary Francis Cusack, The Speeches and Public Letters of the Liberator: With Preface and Historical Notes (Ireland: 1875), 144.

This Royal Commission was significant in that it came from such a high level of government, but it was only one among 114 commissions and 61 committees launched between 1801 and 1845 which inquired into the state of Ireland. Among virtually all these British investigations, the same results could be found: that Ireland was on the verge of starvation with decreasing population numbers, widespread unemployment, and abysmally low standards of living.[223] The reason why the Irish were so reliant on the potato was because landlords had subdivided land holdings, shrinking tenant farms so much that potatoes became the only crop that families could actually live off of. Other crops were usually exported to Great Britain. When blight reached Ireland and the famine began, the British Government did nothing to stop the export of massive amounts of food from Ireland to Great Britain, nor did it lower the price of food in Ireland such that the poor could afford it. According to the tenants of laissez-faire capitalism, the government believed the issue would fix itself. By the time the famine ended in 1849, about one million people had died and another million had emigrated, resulting in an overall population decrease of about 25%. Whether or not it is possible that the famine could have been outright avoided, the fact remains the effects of the famine were exacerbated by a hands-off approach on the part of the people in the best position to help the Irish.

The deadliest famine in recorded history is the Great Chinese Famine of 1959-1961. Estimates of the death toll vary, but usually fall somewhere in the realm of 15-40 million people. This famine came as a result of misguided policies and regulations put forward by the Chinese government as well as flooding and drought in certain regions of China. These policies were part of the Great Leap Forward, a five-year plan launched by the Chinese Communist Party to socialize the country's agriculture and establish China as an industrialized communist society. As a part of this initiative, Chairman Mao Zedong ordered millions of farmers to switch to the production of steel, and all private food production was replaced with smaller-scale agricultural communes, plunging the country's grain production.[224] The Four Pests Campaign was launched to encourage the eradication of four pests: rats, flies, mosquitoes, and most notably, sparrows. Sparrows were framed as a crop seed eating scourge, and were nearly hunted to extinction. As it turned out, the sparrows were also responsible for keeping insect populations in check, leading to the destruction of crops by locusts and other crop-eating insects. Even while grain harvests plummeted, the government published fabricated reports stating that communal

[223] Cecil Woodham-Smith, The Great Hunger: Ireland 1845-1849 (Penguin, 1991): 36.

farming had been successful in bringing grain harvest to an all-time high. Similar to the Irisih Potato Famine, food exports continued even as grain stocks dwindled. These short-sighted efforts to rapidly socialize and industrialize the country went almost completely unnaposed, thanks to the state-owned media controlling the spread of information and the lack of any official opposition party to question the movements of the Communist Party. For a famine this large in scale to be caused largely by irrational policymaking and short-sighted leadership is almost unbelievable. In the end, the causes behind man-made disasters can be just as irrational as the human beings responsible for them.

What if the kind of wrong-minded actions the British and Chinese governments used which inadvertently led to famine (or worsened the famine situation) were purposefully applied by a governing body knowing that they would cause widespread famine? It is this form of genocide that can be seen in the Holodomor. Between 1932 and 1933, around 4 million Ukrainians died from starvation as a result of policies and orders made by the Soviet government under Joseph Stalin. Some of the actions taken include forced collectivization of farms, restricting the movement of people within affected zones, continued export of grain out of Ukraine, confiscation of foodstuffs from the populace, stopping knowledge of the famine from leaving Ukraine, and the refusal to provide any aid for the starving population, among others. It is of historical debate as to whether or not the policies were applied chiefly in order to quell Ukrainian nationalism (which would allow the Holodomor to be legally defined as genocide) or simply to industrialize the country without regard to the wellbeing of the population while making out with as much grain as possible.

[224] Smil, V. "China's great famine: 40 years later." BMJ (Clinical research ed.) vol. 319,7225 (1999): 1619-21. doi:10.1136/bmj.319.7225.1619

Looking back on these past famines it's easy to wonder how such pointless and seemingly avoidable events could have been allowed to occur. The sad truth is that hundreds of millions of people around the world live in near-starvation every single day. Remember the definition of famine: "Lack of access to food to eat." The Food and Agriculture Organization of the United Nations states that over 820 million people, or roughly roughly every one in eight people on earth, live in some form of undernourishment. Research has shown, however, that the world collectively produces enough food for roughly one and a half times the world's current population. Obviously, solving world hunger is not as simple as moving excess food to countries that need it, but it is nonetheless important to recognize modern starvation not as something that is endemic to certain parts of the world but as a famine on a global scale. By looking at the eradication of hunger as something which all nations are responsible for addressing, we pave the way for a future in which global hunger is viewed in the same way as the Irish Potato Famine.

X.
FAMINES AND CHRISTIANITY

> *In everything I did, I showed you that by this kind of hard work we must help the weak, remembering the words the Lord Jesus himself said: It is more blessed to give than to receive.*"[225]

It has always been a great tragedy that hunger exists in a world of plenty. As we have seen in the previous chapter, the effects of famines are devastating. Up until the nineteenth century, the famines which have wiped out whole populations were more often due to natural disasters; however, most cases now are said to be man-made. Not only do famines wreak havoc on entire communities, they also leave behind profound cultural, psychological, religious, social, and spiritual traumas. We are better equipped than ever before to deal with the problem of hunger; thus, the occurrence of famines in the modern world can be said to be a veritable dishonour to humanity.[226] In this chapter, I will briefly discuss the effects of famines in the ancient world, and the cultural, social, and religious effects of more recent famines in various communities around the world, specifically the Irish Potato famine, the Northern China famine, and the Holodomor.[227] Then I will provide a quick summary of the Church's response to famines and the problem of world hunger.

Famines were not uncommon in the ancient world. We can even find multiple accounts of famines occurring in the Bible.[228] Famines often led to a disrupted social hier-

---

[225] Acts 20: 35.

[226] Catholic Church. Pontificium Consilium "Cor Unum." World Hunger, A Challenge for All: Development in Solidarity. 1996. (5).

[227] Holodomor is also known as the Ukrainian Genocide of 1932–33, or the Terror-Famine.

[228] See Genesis 12:10, 26:1,41:53-57, Jeremiah 52:6, Ruth 1:1, 2 Samuel 21:1, etc.

archy as the rich often managed to escape the effects of famine, whereas the poor often became even poorer and more hungry. This led to many riots and serious political consequences as a result. Individuals in ancient communities appeared to have a strong fear of hunger and famine and this would often lead them to do things that they would not normally do. For instance, a farmer who would normally give up when he is confronted by a piece of land which is hard to cultivate would rather continue trying to cultivate the land because of his fear of impending famine. Communal living tended to help with food scarcity and economical hardship as it provided a way to spread food evenly among people living in the same household. Monasteries had a lot of success with preventing hunger within their communities as it was much more cost efficient for them to live together and eat at one table than to live separately on their own. John Chrysostom (347-407), an Early Church Father and bishop of Constantinople, has been known to mention that he could not think of a single case where a monk had died of hunger during his time. Even as far back as the Early Church (around 1-400), the rich were often held responsible for the problem of hunger in their communities as there existed an intentional chasm between the rich and the poor in society. In one of his accounts, Chrysostom speaks of a drought that once overtook his city:

> "PEOPLE PRAYED AND WHEN IT SEEMED GOOD TO GOD, HE SENT PLENTIFUL RAIN. EVERYBODY WAS IN A FESTIVE MOOD, EXCEPT ONE RICH MAN. [THE RICH MAN] SAID THAT HE HAD TEN THOUSAND MEASURES OF WHEAT AND THAT HE WOULD NOW NO LONGER BE ABLE TO DISPOSE OF THEM. WHILST HUNGRY PEOPLE WERE REJOICING BECAUSE OF THE RAIN, THIS RICH MAN GRIEVED."[229]

This displays not only the overwhelming material inequality that exists between the rich and the poor, it

[229] Stander, Hennie. (2010). Chrysostom on hunger and famine. HTS Theological Studies. 67. 10.4102/hts.v67i1.880.

also represents the cruelty that the poor often experience because of it. The inequality grew even greater as hunger and the fear of famine led to a further dependency upon the rich, an increased rate of prostitution, and an increased amount of individuals in prison usually due to acts of desperation.[230] Although the rich may have been held accountable for the various inequalities present in society, the existence of drought, famines, and plagues were commonly thought to have been permitted by God as a way to punish those who have committed grave sins. This idea of God punishing individuals or communities for the sins that they have committed is not exclusive to the ancient world. As we will see further on in the chapter, this idea is, unfortunately, present even in our modern era.

The Irish Potato famine (1845-1849) affected the cultural, religious, and social consciousness of the Irish people in a profound way. Many families and whole communities who had relied on potatoes prior to 1845 were wiped out. The majority of people lost during the famine were Catholic, with the exception of individuals living in parts of east Ulster.[231] The famine was widely regarded as a judgment of God, and although the Catholic Church failed to respond appropriately to the famine, it appears that the position of the Church within Ireland was strengthened, especially during the post-famine decades.[232] The distress caused from the famine triggered a new Reformation, sometimes referred to as "the Protestant Crusade", that encouraged an aggressive evangelization that on many occasions rejected Roman Catholic teachings and traditions. There was an intensified effort to Protestantise the Irish peasants since many individuals, like Sir Charles Trevelyan interpreted the scarcity of produce as:

"[A] JUDGMENT [FROM] GOD ON AN INDOLENT AND UNSELF-RELIANT PEOPLE, AND AS GOD HAD SENT

---

[230] Ibid.

[231] Kinealy, Christine. 1997. A Death-Dealing Famine: The Great Hunger in Ireland. London: Pluto Press. p. 152.

[232] Ibid.

THE CALAMITY TO TEACH THE IRISH A LESSON, THAT CALAMITY MUST NOT BE TOO MUCH MITIGATED: THE SELFISH AND INDOLENT MUST LEARN THEIR LESSON SO THAT A NEW AND IMPROVED STATE OF AFFAIRS [CAN]ARISE."[233]

The Roman Catholic clergy were blamed for the famine as the crisis was largely understood as a necessary intervention from God to change the sinful hearts of the Irish people. The crisis was perceived as a "new sound of [divine] truth to frighten away the devices of [Catholic] falsehood and superstition."[234] Thus, the famine influenced the majority of the Irish Catholic Church due to the overwhelming social dissonance that was produced because of it.[235] Contemporary accounts at the time reveal that natural disasters and misfortunes were commonly viewed as necessary features of life as these events were believed to be used by God to spiritually cleanse His people.

"WHETHER INTERPRETED IN A CHRISTIAN OR A NON-CHRISTIAN FORM . . . THE FAMINE TO THE IRISH PEOPLE WAS 'BEYOND INDIGNATION' AND WAS PERCEIVED AS AN EVENT OF COSMIC SIGNIFICANCE, NOT AS A [MERE] HUMAN CONSPIRACY AGAINST THE IRISH PEOPLE."[236]

Although there were many interpretations of the famine that led to a negative view of the Roman Catholic Church, the position of the Church within Ireland strengthened due to the Church's shift towards a more orthodox and disciplined approach to the Christian faith.[237] The Church became more authoritarian in its approach to religious practices, imposing a new social discipline that spread devotional conformity as a means to amend the cultural impoverishment of Ireland during the post-famine era.[238] The growing influence of the Church also led to the growth of a more patriarchal society than had existed previously.[239] The Northern China famine (1897-1901) resulted in millions

[233] Jenifer Hart, Sir Charles Trevelyan at the Treasury in English Historical Review lxxv, 294, p. 99.

[234] Dunlop, Robert. The Famine Crisis: Theological Interpretations and Implications, p. 167.

[235] It is interesting to note, however, that prior to the famine occurring, only 40 percent of the Irish Catholic population attended church. In 1845, Irish churches had one priest for every 3000 parishioners (see footnote no. 64, no. 66, Ibid., p. 168). Thus, it is difficult to determine just how much the famine negatively impacted the Irish Catholic Church as it appears that they were already in a state of panic prior to the crisis developing.

[236] D.H. Akenson. Small Differences - Irish Catholics and Irish Protestants 1815-1922, p. 144-5.

[237] This was likely a response to the criticisms that the Church was receiving at the time.

238 Kinealy, Christine. 1997. A Death-Dealing Famine: The Great Hunger in Ireland. London: Pluto Press. p. 152.

of deaths and had a significant impact on how the Chinese viewed the Western world. Contemporary missionaries at the time discuss how communities in northern China had interpreted the natural disasters that led to the numerous famines in the area. These communities were convinced that the presence:

> "[OF] SO MANY ARROGANT FOREIGN MISSIONS, CHURCHES AND CATHEDRALS HAD DISRUPTED THE FENG-SHUI OR THE GEOMANTIC BALANCE OF NATURE, THUS AWAKENING THE EARTH DRAGON . . . CAUSING FLOODS AND DROUGHT. AS BOXER "BIG CHARACTER" POSTERS DECLAIMED FROM THE WALLS OF BEIJING: NO RAIN COMES FROM HEAVEN. THE EARTH IS PARCHED AND DRY. AND ALL BECAUSE THE CHURCHES HAVE BOTTLED UP THE SKY."[240]

Corresponding to what we have seen in Chapter 5 with the scapegoating of the Jews during the Black Death, the presence of foreigners, Western philosophies and religions – particularly Christian missionaries – in combination with the numerous droughts that were taking place led many communities in northern China to interpret what was occurring as coming from "a single, occult evil" – the West.[241] This fashioned local sparks of anti-foreignism into a vast populist conflagration across northern China. As the presence of Christian missionaries increased between 1890 to 1908, Westerners became gradually seen as "foreign devils."[242] The locals viewed the construction of churches and the presence of missionaries as a "religious invasion" that intended to dismantle the Chinese way of life.[243] As hunger became more prevalent, rumours spread of Christians poisoning wells and foreign ships being discovered to contain human body parts.

As a result of the increasing fear towards Western foreigners, the anti-Christian "Spirit Boxers" movement

---

[239] Ibid., 153. This led to many Irish women seeking opportunities outside of the country.

[240] Davis, Mike. Late Victorian Holocausts: El Nino Famines and the Making of the Third World. New York: Verso, 2001, p. 179.

[241] Ibid.

[242] Ibid., 179-181.

[243] Ibid., 181.

developed and spread across the country, particularly in areas where harvests had failed. The Spirit Boxers violently seized food from neighboring Christian missions and villages commanding villagers to give them all of their grain and produce. If the targeted communities did not give them what they wanted, then the Spirit Boxers would take the food by force. It is interesting to note that while a China Famine Relief Fund was being organized in London, China missionary supporters viewed "[this] famine relief as a heaven-sent opportunity to spread the gospel."[244] Although there were many Christian missionaries sent to northern China, they were not successful in their evangelization. They often encountered hostility with the locals due to the anti-foreignism that had cultivated in the land. Converts to Christianity, frequently termed "rice Christians", were short-lived, gradually causing missionaries to lose evangelical momentum.

Holodomor (1932-1933) had killed around 4 million Ukrainian people. The term 'Holodomor', is a compound word that refers to moryty holodom – the Ukrainian phrase used to explain how someone is starved to death, indicating that the famine was planned and organized.[245] Horror, indifference, exhaustion, starvation, and violence became a common experience for millions of Ukrainian people due to the Stalinist regime. As Anne Applebaum states in Red Famine: Stalin's War on Ukraine:

> "FACED WITH TERRIBLE CHOICES, MANY MADE DECISIONS OF A KIND THEY WOULD NOT PREVIOUSLY HAVE BEEN ABLE TO IMAGINE. ONE WOMAN TOLD HER VILLAGE THAT WHILE SHE WOULD ALWAYS BE ABLE TO GIVE BIRTH TO OTHER CHILDREN, SHE HAD ONLY ONE HUSBAND, AND SHE WANTED HIM TO SURVIVE. SHE DULY CONFISCATED THE BREAD HER CHILDREN RECEIVED AT A LOCAL KINDERGARTEN, AND ALL HER CHILDREN DIED.

[244] Ibid., 77.
[245] Stähle, Hanna. Zeitschrift Für Slavische Philologie 70 no. 2 (2014): 457-62. Accessed July 1, 2020. www.jstor.org/stable/43974109.

A COUPLE PUT THEIR CHILDREN IN A DEEP HOLE AND LEFT THEM THERE, IN ORDER NOT TO HAVE TO WATCH THEM DIE.[246]

In these dire circumstances, there was no point in keeping the rules of ordinary morality. As the need for self-preservation intensified, empathy towards one's neighbor became increasingly discouraged. The famine led to extraordinary acts of aggression and theft. People stole whatever food they could find. Due to extreme desperation and hunger, many were killed for harbouring food while some took up cannibalism and consumed the dead. There were lynch mobs that brutally tortured individuals for stealing crumbs of bread. Those who were seriously ill were often buried alive. Children were hunted down as food. Human meat markets grew in size and number. Indifference had begun to spread at the same rate that people died.

For many historians of Ukraine, the Holodomor was not only a genocide in the sense that Stalin's regime killed millions of people, it was also a genocide because these individuals were killed with the very purpose of destroying the Ukrainian way of life.[247]

Just as millions of people were dying, the Soviet authorities waged war against "Ukrainian nationalism" with a large-scale assault on the Ukrainian church. This radically changed the Ukrainian religious way of life as it led to the material impoverishment of many villagers.[248]

This attack on cultural and religious traditions ultimately led to the destruction of many religious practices and institutions. Religious practices associated with the church, i.e., baptisms, weddings, funerals, and religious holidays, were special targets as these were central events for Ukrainian country folk.[249]

---

[246] Applebaum, Anne. Red Famine: Stalin's War on Ukraine, 2017, p. 160.

[247] Irvin-Erickson, Douglas. "Genocide, the 'Family of Mind' and the Romantic Signature of Raphael Lemkin." Journal of Genocide Research 15.3, 2013. p. 273-96.

[248] Andriewsky, Olga. "Towards a Decentred History: The Study of the Holodomor and Ukrainian Historiography." East/West: Journal of Ukrainian Studies, 2(1), 2015. pp. 36-40.

[249] Andriewsky, Olga. "Towards a Decentred History: The Study of the Holodomor and Ukrainian Historiography." p. 40.

For instance, traditional Ukrainian funeral rites i.e., the singing of psalms, reading from the Bible, and the use of professional mourners, were prohibited. No energy was to be wasted digging graves, holding ceremonies, or playing music.

> "FOR A CULTURE THAT HAD VALUED ITS RITUALS HIGHLY, THE IMPOSSIBILITY OF SAYING A PROPER FAREWELL TO THE DEAD BECAME ANOTHER SOURCE OF TRAUMA: THERE WERE NO FUNERALS, RECALLED KATERYNA MARCHENKO. THERE WERE NO PRIESTS, REQUIEMS, TEARS. THERE WAS NO STRENGTH TO CRY."[250]

Almost every aspect of the Ukrainian way of life was destroyed during the Holodomor era. Even seemingly innocent traditional practices like the *dosvitky*[251] were suppressed. The very foundations for Ukrainian religious and social organizations were demolished, extinguishing any remnants of Ukrainian autonomy. Although there was a lot of fear cultivating in the lands, there were some Ukrainian families who received food and drink from their Jewish neighbours. Jews were often spared from the effects of the famine because they were not farmers and, therefore, did not own any land. These families were often saved due to these benevolent gestures. "At a time when hatred and suspicion of all kinds were rising," Anne Applebaum states, "the gesture was a powerful one."[252] In 1933, a debate broke out inside the Vatican regarding the Ukrainian famine. There was one group that wanted to send a famine relief mission to the USSR, and another that opted for diplomatic caution.[253] The latter group won, and although the Vatican was continually receiving information about the famine, they had decided to keep quiet about what was occurring in Ukraine to the public. The Holy See decided to remain silent for a multitude of reasons:

[250] Applebaum, Anne. Red Famine: Stalin's War on Ukraine, p. 164.
[251] The "dosvitky" was a Ukrainian winter evening ritual where young, unmarried women often gathered to sew, embroider, and sing.
[252] Ibid., 172.
[253] Ibid., 195.

"[S]UCH AN EFFORT WOULD INVOLVE OTHER FAITH GROUPS; THE VATICAN ITSELF WAS IN FINANCIAL CRISIS; THEIR EFFORTS WOULD [HAVE BEEN] REBUFFED BY SOVIET AUTHORITIES, WHO DENIED THE EXISTENCE OF THE FAMINE, AND ANY ASSISTANCE DELIVERED WOULD LIKELY NOT END UP IN THE HANDS OF THOSE WHO NEEDED IT; THE VATICAN'S EXPOSURE OF THE FAMINE WOULD [HAVE] LIKELY RESULT[ED] IN INCREASED SOVIET PERSECUTION OF CATHOLICS; AND THEIR EFFORTS WOULD [HAVE] PLAY[ED] INTO THE HANDS OF NAZI ANTI-SOVIET PROPAGANDA."[254]

Pope Pius XI opted for diplomatic caution as it appeared that it would not have been helpful to the Ukrainian people to sponsor a famine relief mission when the Soviet regime refused to even acknowledge the devastation occurring in the Ukraine. The Vatican's decision was strongly influenced by the rise of the Nazis to power in Germany, whose anti-Soviet propaganda took every advantage of the famine to condemn Stalin and his Communist policies.[255] Pope Pius XI's major priority at the time was to secure the Nazi government's agreement to a Reichskonkordat, a treaty negotiation between the Vatican and Nazi Germany, that served to protect Catholic Church members and institutions by a recognition of the principal rights of the Church.[256] Therefore, any acts which could be interpreted as assisting the Soviet Union might have had fatal consequences for the Catholics living in Nazi Germany. Although the Vatican attempted to draw awareness through L'Osservatore Romano, the Vatican newspaper, and provided limited financial assistance through indirect channels to the Ukraine, the Vatican's non-interventionist approach was a disappointment to many that expected the Church to live up to its moral professions, i.e., helping others who are in crisis. This is especially due to the fact that the Nazis still violated the treaty by shutting down Catholic institutions, confiscating Church property,

[254] Kenworthy, Scott. "The Holy See and the Holodomor. Documents from the Vatican Secret Archives on the Great Famine of 1932-1933 in Soviet Ukraine. Toronto: Kashtan Press/University of Toronto, Chair of Ukrainian Studies, 2011. 978 1 896354 37 8." The Journal of Ecclesiastical History 65, no. 1 (2014): 228-28. doi: 10.1017/S0022046913001127.

[255] Book Note: A. D. McVay and L. Y. Luciuk, eds., The Holy See and the Holodomor. Documents from the Vatican Secret Archives on the Great Famine of 1932-33 in Soviet Ukraine (Toronto: University of Toronto and Kashtan Press, 2011), 99 Pp., ISBN 978-1-896354-37-8.

[256] "An Agreement with the Catholic Church," Facing History and Ourselves, accessed July 9, 2020, https://www.facinghistory.org/holocaust-and-human-behavior/chapter-5/agreement-catho-

and imprisoning clergy and other Catholic leaders.[257]

As we have seen in this chapter, famines often reveal to us questionable aspects of human nature. The scarcity of food often produces a lack of empathy for others as the primal instinct to survive takes over. During these times, it is rare to find instances of the Christian vocation to love being lived out, and this is understandably so. We have seen how some Christians have used times of famine as a means of evangelization. Whether or not this evangelical approach is an appropriate response to pandemics remains a question that many are struggling to answer to this day. Regardless of how others have responded to these unfortunate circumstances, Christians have the moral obligation to help those in need, particularly those who are struggling with having their basic needs met., i.e., food, water, shelter, freedom, etc. As revealed in Sacred Scripture, we find that the vocation to love not only demands a sharing of one's riches with the poor, it also describes a necessary interior transformation in order to view others in the world as one's spiritual family:

> "THE HEART OF THE POOR LAZARUS WAS FREER THAN THE HEART OF THE EVIL RICH MAN, AND THROUGH THE VOICE OF ABRAHAM, GOD NOT ONLY ASKS THE UNRIGHTEOUS RICH MAN TO SHARE HIS FEAST WITH LAZARUS, BUT DEMANDS A CHANGE OF HEART AND ACCEPTANCE OF THE LAW OF LOVE, IN ORDER THAT HE BECOME A BROTHER TO LAZARUS (CF. LK. 16:19 FF)." [258]

The consequences of famines are devastating. When people are not able to look at others with dignity and love, then it is not difficult to find individuals and whole communities being used as a means to an end. The Christian vocation to love demands that individuals see others as they are, as beings made in the image and likeness of God. Viewing others in this way makes it possible to love them rightly; and when this occurs, individuals cannot help but offer assistance and support to those who are poor, hungry, and suffering.

lic-church.
[257] Ibid.
[258] Catholic Church. Pontificium Consilium "Cor Unum." World Hunger, A Challenge for All: Development in Solidarity. 1996. (65).

XI.
MODERN CHRISTIAN THEOLOGY

> "Respect and love ought to be extended also to those who think or act differently than we do in social, political and even religious matters. In fact, the more deeply we come to understand their ways of thinking through such courtesy and love, the more easily will we be able to enter into dialogue with them... God alone is the judge and searcher of hearts, for that reason He forbids us to make judgments about the internal guilt of anyone."[259]

Although modern Christian theology is immensely varied, the main aim of this chapter is to introduce the themes commonly found within contemporary Church theology, i.e., love, inclusivity, and modernity. This chapter will briefly discuss the neoscholasticism that was prevalent prior to the Second Vatican Council, and the influences that the nouvelle théologie movement had on the Church's overall rejection of neoscholasticism in theology. Then, we will look at contemporary Church documents on the Christian vocation to love and serve others in the world, particularly those who are suffering or living in poverty.

Neoscholasticism is a branch of modern Christian theology that emphasizes the institutional structures of the Church using a more manualist method

---

[259] Vatican Council. 1998. Pastoral constitution on the Church in the modern world: Gaudium et spes ; promulgated by His Holiness Pope Paul VI on December 7, 1965. (28).

[260] When I speak of modernity, I am largely referring to the time after the end of World War I (1914-18) to our present time. For the purposes of this chapter, I will also be using the term "modernity" to include the postmodern age (1980s-present).

[261] It is interesting to note that many philosophers and theologians have viewed the neoscholasticism at the time as being overly reductive in its presentation of Aquinian metaphysics and theology.

[262] This was due to Aquinas'

of understanding the Christian faith and the world. A defining trait of neoscholasticism is its avoidance of modernity.[260] Neoscholasticism is an intellectual system which is thought to have been conceived in the Middle Ages, particularly in the writings of St. Thomas Aquinas.[261] Neoscholastic theology tends towards meticulously defined concepts, questions, and answers; thus, the range of acceptable theological responses are often restrictive in nature. Responses to theological questions needed to be done with an intense precision and clarity.[262] Although it was not uncommon to find neoscholastic authors using the history of the early Church and Sacred Scripture for their various theological propositions, historical context was often neglected or outright disregarded. Instead, these authors depended primarily on the application of deductive logic to demonstrate various theological truths.

By the second half of the nineteenth century, neoscholasticism was established as the accepted form of Roman Catholic thought with many seminaries teaching theology in this manner.[263] The neoscholastic approach had profoundly changed Catholic theology by the early part of the twentieth century, emphasizing still the deductive nature and precision of Aristotelian metaphysics. With this emphasis on obligatory rules and regulations, the holiness of the Church was equated with the obedience of her members. Thus, members of the Church were commonly viewed as "merely rule-followers and not full, active participants in the life of the Church."[264] By the end of World War I, however, theologians, particularly in France and Germany, began to question the role that neoscholasticism played in the West. They were concerned that the neoscholastic method would gradually fashion an impenetrable barrier between the Church and others in the secular world. The manualist approach towards teaching Church

adoption of Aristotelian logic and metaphysics. You could compare neoscholasticism to the early 20th century analytic movement in philosophy as both emphasize the significance of conceptual clarity by applying various logical or systematic methods. This striving towards greater conceptual precision is consistent with the methods used in the modern sciences.

[263] Ford, David F. Modern Theologians: An Introduction to Christian Theology Since 1918, 2012. p. 94

[264] Gribaudo, Jeanmarie. "Three Key Moments in the Developing Theology of the Holiness and Sinfulness of the Church in the Twentieth Century". Sacred Theology Degree, Boston College, 2012 p. 12.

doctrine and theology was seen as "[overly] defensive and apologetic."[265] Considering the social and political turmoil at the time,[266] these theologians proposed a new, more Scripturally grounded way of doing theology. This new way of doing theology emphasized both the historical and scriptural dimensions of the Church, focusing on the application of Sacred Scripture and Church Traditions to the theological and existential questions being posed at the time. They found that focusing on liturgical, patristic, and scriptural studies helped to improve the relationship between the lived, ecclesial[267] dimensions of faith and theology with the more cognitive and philosophical pursuits so prominent in neoscholasticism.[268] This new movement in theology is characterized by its propensity towards ressourcement (a returning to the sources) and eventually led to the rediscovery of Patristic thought. From this, the nouvelle théologie (new theology) movement came into existence. It is thought to have its origins in the monastery Le Saulchoir of France, beginning around 1935.[269] The term originated from a 1942 article in L'Osservatore Romano which intended to condemn the movement, claiming that its focus on individual experience, subjectivity, and religious sentiment, were highly exaggerated and needed to be rejected.[270] However, with Pope John XXIII's calling of the Second Vatican Council (1962-65), there was a:

> "...FAVOR [TOWARDS] DOCUMENTS MORE SCRIPTURALLY ROOTED AND PASTORAL IN TONE, [GIVING] OFFICIAL SANCTION TO THE MOVE BEYOND NEOSCHOLASTICISM AS THE ONLY VALID MODE OF CATHOLIC THEOLOGICAL REFLECTION. AT WORK IN THE RESULTING DOCUMENTS, PREEMINENTLY SO IN GAUDIUM ET SPES, IS A RECOVERY OF THE GRAND VISION OF CATHOLIC THEOLOGY AS CONCERNED TO UNDERSTAND THE SIGNIFICANCE OF ALL PARTICULAR THINGS IN RELATION TO GOD'S SELF-REVELATION IN CHRIST AND AS CORRELATIVELY OPEN TO THE

---

[265] Ibid., 3.

Christian apologetics is a branch of Christian theology that focuses on defending the faith against other religious or secular objections.

[266] The Great Depression and the two world wars.

[267] Ekklesia, or the Church, is a term often used in systematic theology to indicate the community of believers that have identified themselves as being inspired by the Word of God revealed in history, Sacred Scripture, Tradition, and in/through the Spirit.

[268] Ford, David F. Modern Theologians: An Introduction to Christian Theology Since 1918, 2012. p. 271.

[269] Gribaudo, Jeanmarie. "Three Key Moments in the Developing Theology of the Holiness and Sinfulness of the Church in the Twentieth Century". p. 3.

[270] Ibid., 4.

CONTRIBUTIONS OF THE DIVERSE FORMS OF ANALYSIS THIS REQUIRES. IT IS THIS VISION OF ROOTED, PRACTICALLY ENGAGED, DISCIPLINARY PLURIFORM THEOLOGICAL ANALYSIS AND REFLECTION THAT [CONTINUES] TO CHARACTERIZE CATHOLIC THEOLOGY AFTER VATICAN II."[271]

The decision to move beyond a neoscholastic approach to theology was ultimately due to the growing desire for unity within the Church. This striving towards Church unity is evident in as far back as the Patristic Period (100-451) with theologians, such as St. Augustine, arguing for a unity that models the divine and infinite love of Christ on the cross. This demanded a model of unity that "[was] broad enough and human enough to include a multitude of sinners", referencing the inclusivity of love that Christ demonstrated with His self-sacrificial act in order to save humanity from sin (or, in other words, death).[272] For the purposes of this chapter, I will be drawing upon several twentieth-century theologians such as, Henri de Lubac, and Hans Urs von Balthasar, to demonstrate the Church's efforts to move beyond mere neoscholastic thought, emphasizing instead the communion-based, ecclesial vision of the Second Vatican Council. Associated with the nouvelle théologie movement, these theologians reintroduced the Patristic idea that in order to be Catholic, one needed to be inclusive, as well. As Dennis Doyle states in *Communion Ecclesiology*:

> "TO BE "INCLUSIVE" HERE DOES NOT MEAN THAT NO ONE CAN EVER BE EXCLUDED; RATHER, IT MEANS PARADOXICALLY, THAT THE ONLY REASON TO EXCLUDE SOMEONE WOULD ULTIMATELY BE FOR THEIR OWN LACK OF INCLUSIVITY. THE INTENT IS NOT TO MARGINALIZE PEOPLE OR VIEWS, BUT SIMPLY TO ACKNOWLEDGE THAT CERTAIN POSITIONS THEMSELVES MARGINALIZE WHAT SHOULD BE CENTRAL. THE CATHOLIC IMPULSE IS TO FAVOR THE "BOTH/AND" OVER

[271] Ford, David F. Modern Theologians: An Introduction to Christian Theology Since 1918, 2012. p. 271-72.

[272] Gribaudo, Jeanmarie. "Three Key Moments in the Developing Theology of the Holiness and Sinfulness of the Church in the Twentieth Century". p. 28.

> THE "EITHER/OR." [TO BE CATHOLIC] IS TO BE OPEN TO TRUTH WHEREVER IT MAY BE FOUND. IT IS TO OPT FOR UNITY, SOMETIMES AT THE COST OF OTHER GOODS."[273]

This reorientation towards inclusivity signaled a rejection of the restrictive and often rigid nature of neoscholastic theology. The ecclesial vision of the Council strived to move beyond mere juridical and institutional understandings of the Church; emphasizing instead, the historical, mystical, and sacramental dimensions. This cultivated a greater understanding of how Christians ought to be in their relationships with one another, with individuals in secular society, and the world.

The French Jesuit theologian, Henri de Lubac (1896-1991), was considered by many to be a master of paradox. De Lubac emphasized the importance of the principle of *complexio oppositorum*: "the dynamic holding in tension of contrary points."[274] With this principle, de Lubac reminds Christians that the gospels are full of paradox, and that spiritual truth is in itself of a paradoxical nature; for instance, the nature of freedom with its dependency on grace, the relationship between unity and diversity, and the divine yet human nature of Jesus Christ.[275] De Lubac was unlike the neoscholastic writers at the time as he embraced the existence and reality of mystery wholeheartedly. Since the neoscholastic approach emphasized clarity and precision, de Lubac thought that it was ill-suited for the study of theology because it was unwilling to properly encounter the reality of mystery. Mysteries cannot be precisely formulated, nor can they be completely understood, and since mysteries exist in the very character of Christian life and the Church, there would be a significant amount missing from a neoscholastic method of doing theology.

---

[273] Doyle, Dennis. Communion Ecclesiology: Visions and Versions. Maryknoll: Orbis Books, 2000. p. 21.

[274] Doyle, Dennis. Communion Ecclesiology: Visions and Versions. p. 100.

Throughout its history, various imagery has been used to describe the Church. Images such as, "the Bride of Christ, "people of God", "temple of God", and "the Body of Christ", serve to instill a divine relational existence within the historical-temporal dimension that humanity resides in. There exists a great mystery within the character of the Church as a result of the unification of the vertical (or divine) dimension with its horizontal (or historical-temporal) component. Since it is through the work of the Holy Spirit that the Church "receives her consistency as a people", this means that the Church is dependent upon the Spirit (the vertical, divine component) to understand her mission in the world.[276] This mystery within the character of the Church necessitates the Catholic disposition to embrace the spiritual mysteries and truths that are at times expressed in paradox. It is a refusal to reduce mystery to anything manageably one-sided and partial.[277] Catholicity for de Lubac, then, is: "not only an encompassing of various dimensions of truth held in tension, and not only a socially conscious embrace of all that is good and worthy, but also a radical inclusion of all human beings in all of their depth and mystery."[278] You may begin to wonder: Just what does it mean to accept or embrace someone in all of their depth and mystery? In order to demonstrate what this looks like, I will use a brief analogy. Let us imagine a man standing in front of a mirror. Upon seeing his reflection, the man is able to identify himself as it is his reflection that he sees in the mirror. I think we would all agree that there is nothing wrong with him identifying his reflection in the mirror. However, if he, instead, equated his reflection in the mirror with his very existence, then we would almost certainly have to disagree with him. Now, why would this be the case? We would disagree with him because there is so much more to the man than what he sees in the mirror; he is

[275] Ibid., 58-61.

[276] International Theological Commission: vol. I: Texts and Documents (1969-1985), pref. J. Ratzinger, ed. M. Sharkey, Ignatius Press, San Francisco 1989, p. III.4.

[277] Doyle, Dennis. Communion Ecclesiology: Visions and Versions. p. 61.

[278] Ibid., 62.

beyond a mere reflection of himself. This simple analogy demonstrates what it is like when we do not embrace the depth or mystery of another individual. Instead of taking into consideration what we do not know about an individual, i.e., his circumstances, history, and personal life, we would rather take what we can observe, i.e., his appearance, behavior, and statements, and falsely equate it with his very identity.

To embrace the depth and mystery of another individual is to explore the meanings of things beyond their surface appearances. It requires an open willingness to discover the depths of another person's interior life, i.e., their fears, hopes, joys, and sorrows. It is to move beyond superficial things, i.e., the colour of their skin, the quality of their clothing, or their profile on social media. Thus, to fathom the mystery of another person is to be aware that there is so much more to them than what they can ever reveal to us or what we can ever perceive. This disposition of humility and openness to mystery is what it means to be Christian; therefore, this orientation towards mystery ought to be incorporated into a Christian theology.

The Swiss theologian, Hans Urs von Balthasar, strived to encounter the reality of mystery within his systematic theology of the human condition.[279] He pursued theological studies at the Jesuit theologate at Lyon in France where he encountered neoscholasticism's "narrow intellectualism."[280] Balthasar expressed his concerns of a neoscholasticism that reduced the mysteries of human life to a mere set of systems, ideologies, or logical formulas. He emphasized, instead, the mystery of the relational existence of the Creator within the human condition.

"THIS VULNERABLE YET ENDURING RADIANCE AND MYSTERY INHERENT IN ALL REALITY WAS FOR BALTHASAR THE GREAT SIGN OF ITS MIRACULOUS SOURCE — AS

---

[279] It is interesting to note that Balthasar studied under the influence of de Lubac. It was most likely de Lubac who introduced Balthasar to the nouvelle théologie movement.

[280] Gribaudo, Jeanmarie. "Three Key Moments in the Developing Theology of the Holiness and Sinfulness of the Church in the Twentieth Century". p. 57-8.

GIFT, AS FREELY BESTOWED, AS CREATION SPRINGING FROM THE CREATOR. THEOLOGY CAN ONLY BE TRUE INSOFAR AS IT CONTINUALLY RECOLLECTS THIS DEEP SOURCE OF ANY SORT OF TRUTH WHATSOEVER. THE ULTIMATE GROUND OF THE MYSTERIOUS CHARACTER INHERENT IN THE KNOWABLE IS DISCLOSED ONLY WHEN WE RECOGNIZE THAT EVERY POSSIBLE OBJECT OF KNOWLEDGE IS CREATURELY, IN OTHER WORDS, THAT ITS ULTIMATE TRUTH LIES HIDDEN IN THE MIND OF THE CREATOR, WHO ALONE CAN SPEAK THE ETERNAL NAME OF THINGS."[281]

Balthasar focused primarily on theological themes concerning the finiteness of humanity within the infinite nature of God. Influenced by the Ignatian spirituality of discovering God in all existing things, Balthasar expressed how individuals can paradoxically experience the holiness and glory of God through a loving of what is fragile, sinful, and lowly.[282] It is through his Ignatian spirituality that Balthasar connects academic theology to human history, Church Tradition, and spirituality; thus, providing a systematic theology that overcomes the rigid confines of neoscholasticism. As Sister Jeanmarie Gribaudo states:

> "THE BEAUTY OF THE CHURCH IS FOUND IN ITS TIMELESSNESS; IT IS EVER ANCIENT, EVER NEW. YET, BECAUSE THE CHURCH IS A HISTORICAL REALITY, IT CAN AND MUST GO THROUGH "A CONTINUAL STATE OF REBUILDING" AND ADAPT TO THE NEEDS AND PROBLEMS OF THE CURRENT ERA IN WHICH IT IS EXISTING. DOING THEOLOGY IN A WAY THAT ADAPTS TO AND ADDRESSES THE CURRENT MILIEU IS CLEARLY A BREAK FROM THE NEO-SCHOLASTIC METHOD, WHICH WAS A PREDICTABLE, UNCHANGING PRESENTATION OF THESIS, PROOF, AND APPLICATION."[283]

With the move beyond mere neoscholasticism, the Church has discussed many aspects of modernity

---

[281] Ford, David F. Modern Theologians: An Introduction to Christian Theology Since 1918, 2012. p. 403

[282] Gribaudo, Jeanmarie. "Three Key Moments in the Developing Theology of the Holiness and Sinfulness of the Church in the Twentieth Century". p. 60.

[283] Ibid., 47.

[284] Vatican Council. 1965. Dogmatic constitution on the Church: lumen gentium / solemnly promulgated by His Holiness, Pope Paul VI on November 21, 1964. Being "not of this world" indicates the Christian belief that there exists a spiritual component to the human condition.

in her documents and encyclicals. Lumen Gentium (Light of the Nations) is one of the main documents of the Second Vatican Council. It is here that we find the discussion for both the holiness and the sinfulness of the Church, emphasizing the Church's call to serve others in the world because its members are not of it.[284] In Gaudium et Spes (Joy and Hope), Pope Paul VI describes the defining traits of modernity that Christians must consider:

> "NEVER HAS THE HUMAN RACE ENJOYED SUCH AN ABUNDANCE OF WEALTH, RESOURCES AND ECONOMIC POWER, AND YET A HUGE PROPORTION OF THE WORLD'S CITIZENS ARE STILL TORMENTED BY HUNGER AND POVERTY, WHILE COUNTLESS NUMBERS SUFFER FROM TOTAL ILLITERACY. NEVER BEFORE HAS MAN HAD SO KEEN AN UNDERSTANDING OF FREEDOM, YET AT THE SAME TIME NEW FORMS OF SOCIAL AND PSYCHOLOGICAL SLAVERY MAKE THEIR APPEARANCE. ALTHOUGH THE WORLD OF TODAY HAS A VERY VIVID AWARENESS OF ITS UNITY AND OF HOW ONE MAN DEPENDS ON ANOTHER IN NEEDFUL SOLIDARITY, IT IS MOST GRIEVOUSLY TORN INTO OPPOSING CAMPS BY CONFLICTING FORCES. FOR POLITICAL, SOCIAL, ECONOMIC, RACIAL AND IDEOLOGICAL DISPUTES STILL CONTINUE BITTERLY, AND WITH THEM THE PERIL OF A WAR WHICH WOULD REDUCE EVERYTHING TO ASHES. TRUE, THERE IS A GROWING EXCHANGE OF IDEAS, BUT THE VERY WORDS BY WHICH KEY CONCEPTS ARE EXPRESSED TAKE ON QUITE DIFFERENT MEANINGS IN DIVERSE IDEOLOGICAL SYSTEMS."[285]

In Deus Caritas Est (God is Love), Pope Benedict XVI discusses the divine and eternal nature of love that Christians ought to conform themselves to. It is here that the loving role of the Holy Spirit is emphasized as endlessly working in and through individuals, regardless of their cultural, religious, or spiritual backgrounds.[286] Thus, it can

---

[285] Vatican Council. 1998. Pastoral constitution on the Church in the modern world: Gaudium et spes ; promulgated by His Holiness Pope Paul VI on December 7, 1965.

[286] Catholic Church, and Benedict. 2006. Encyclical letter Deus caritas est of the Supreme Pontiff Benedict XVI to the bishops, priests and deacons, men and women religious and all the lay faithful on Christian love.

be said that God is evident in all areas of the modern, secular world. In Laudato Si' (Praise be to You), Pope Francis illustrates the importance of understanding the relational existence that the human condition has with the divine. In this encyclical, he critiques the capitalist materialism that has generated not only an "intensified pace of life and work", but also the increase of pollution that continues to negatively impact people in the world, particularly those who are living in poverty.[287] As we have seen in this chapter, the development of modern Christian theology from neoscholasticism to its present indicates the growing need for action as it deals with the problems of modernity. The Church strives for an inclusivity and unity that displays the very model of love revealed to Christians in the Blessed Trinity. As it states in the opening paragraph of Unitatis Redintegratio (Restoration of Unity): "[D]ivision openly contradicts the will of Christ, scandalizes the world, and damages the holy cause of preaching the Gospel to every creature"; therefore, Christians must, with the grace of the Spirit, pray and act towards loving unity with all members of the human race.[288] This loving call towards unity demands that Christians help others regardless of who they are or where they come from. With the Second Vatican Council, Christians are reminded of the moral obligations that they have towards others in the modern world, particularly those who are suffering with hunger, illness, and poverty.

[287] Catholic Church, and Sean McDonagh. 2016. On care for our common home: the encyclical of Pope Francis on the environment, Laudato Si'.

[288] Cassidy, Edward Idris. 2005. Ecumenism and interreligious dialogue: Unitatis redintegratio, Nostra aetate. New York: Paulist Press.

XII.
HIV/AIDS

> **"WE ARE ALL HUMAN, AND THE HIV/ AIDS EPIDEMIC AFFECTS US ALL IN THE END. IF WE DISCARD THE PEOPLE WHO ARE DYING FROM AIDS, THEN WE CAN NO LONGER CALL OURSELVES PEOPLE. THE TIME TO ACT IS NOW. WE CAN MAKE A DIFFERENCE."**
> **- NELSON MANDELA**[289]

The HIV/AIDS pandemic has been plaguing the world for over a century, and in that time it has been able to spread and infect upwards of 50 million people all across the globe, with some estimates putting the number closer to 100 million.[290] In that 100 year span, more than half of it was spent unaware of the existence of the virus, with efforts to uncover and understand it not beginning until the late 70s when a trend of rare diseases and conditions was noticed amongst gay men who otherwise should have been healthy and able to combat them. Although we spent more than half of its existence unaware of it, in the time since its discovery we've made strides in combatting and treating HIV, to the point where for many it is no longer the death sentence it once was, and those living with HIV can reasonably expect to live as long as anyone else.[291] Although HIV/AIDS is a global concern, the virus hasn't had an equal impact everywhere it's been, and to this day some countries and specific populations still suffer more from and are at a higher risk of becoming infected with HIV. Despite this inequity, the world has taken great strides in addressing, treating, and preventing HIV/AIDS, with

---

[289] Nelson Mandela, "Care, Support, Destigmatization," Closing Statements at the XVI International AIDS Conference Barcelona, Spain, July 12, 2002.

[290] "Global HIV & AIDS statistics - 2020 fact sheet," UNAIDS, 2020, https://www.unaids.org/en/resources/fact-sheet.

[291] "HIV/AIDS," World Health Organization, Jul. 6, 2020, https://www.who.int/news-room/fact-sheets/detail/hiv-aids.

"The Science of HIV and AIDS - Overview," Avert, Oct. 10, 2019, https://www.avert.org/professionals/hiv-science/overview.

many organizations working across the globe to improve access to treatment for those infected and provide education on the virus so that the stigma surrounding it and the spread of the virus itself can be reduced. The aim of this chapter is to provide a general understanding of what exactly HIV is, the history of the virus, and how it was able to spread throughout the world undetected for so long, and information on those groups most at risk and the countries that have been impacted the most by this pandemic.

HIV stands for Human Immunodeficiency Virus, and it is a retrovirus that attacks the immune systems in those infected with it, specifically a type of white blood cells called CD4 T cells that help our body respond to infections.[292] HIV is called a retrovirus because rather than store its genetic information using DNA, it instead stores it through RNA,[293] meaning that once HIV infects a cell, it has to make a DNA version of it before inserting that into the host genome in order to make copies of itself.[294] The virus then spreads through the body through this method, killing the CD4 T cells as it does so, until the point where the infected person's T cell count falls below 200 cells per microliter of blood, where the normal amount is anywhere from 500-1500.[295] It's at this point, or when a person develops an opportunistic infection – infections that someone with an average immune system would be normally be able to fight off – that a person is diagnosed as having AIDS – Acquired Immune Deficiency Syndrome, and it's usually the opportunistic infections that occur at this point that end up killing AIDS patients.[296] There are two distinct types of HIV, referred to as HIV-1 and HIV-2. Of these two viruses, HIV-1 is by far the most common, being responsible for around 95% of all worldwide infections, whereas the majority of cases of HIV-2 are found in West Africa.[297] Additionally, there are four groups of HIV-1 strains, called groups M, N, O, and P.[298] HIV-1 group M is

[292] "The Science of HIV and AIDS - Overview," Avert, Oct. 10, 2019, https://www.avert.org/professionals/hiv-science/overview.
[293] Ibid.
[294] SciShow, "How a Sick Chimp Led to a Global Pandemic: The Rise of HIV," Youtube Video, 10:19, Nov. 29, 2017, https://www.youtube.com/watch?v=izwomieBwG0.
[295] Ibid.
[296] Ibid.

the strain that makes up the vast majority of HIV cases, while the other 3 groups are particularly uncommon. From here, HIV-1 group M can be further broken down into nine different subtypes, named A, B, C, D, F, G, H, J, and K, which can additionally combine their genetic material in order to make hybrid viruses, of which 89 are known to exist.[299] Subtype B is the dominant HIV subtype across the Americas, Australasia, and Western Europe, and has been the subtype researched the most despite only making up 12% of all HIV infections.[300] The HIV subtype that constitutes the greatest number of infections is subtype C, which makes up approximately half of all infections, and is particularly common in countries with high rates of infections. Despite all we know about the HIVs and the various strains and subtypes, we don't know definitely whether the differences in subtypes matter – although we do know that HIV-2 is less infectious and progresses slower than HIV-1 – with some earlier studies suggesting significant differences between the subtypes, but more recent research calling those findings into question.[301] Regardless, the main methods of treating HIV seem to be effective for all of the subtypes and strains, so despite the inequality in the research, there are still viable treatment methods for people living with any type of HIV.

Despite all we know about HIV and AIDS now, back in the 1980s when the early cases began to appear, all that was known was that suddenly healthy young gay men were suddenly dying from extremely rare infections.[302] Between the fall of 1980 and the spring of 1981, an immunologist from the UCLA saw 5 patients, all of which were gay men in their 20s to 30s suffering from a rare form of pneumonia caused by a fungus growing inside their lungs – a fungus that normally would be harmless and incapable of infecting the lungs. By June, two of the men had died.[303] hortly thereafter, a dermatologist from New York noted a similar trend, except this time with a rare form of cancer called

[297] "HIV Strains and Types," Avert, Feb. 26, 2019, https://www.avert.org/professionals/hiv-science/types-strains.
[298] Ibid.
[299] Ibid.
[300] Ibid.
[301] Ibid.
[302] SciShow, "How a Sick Chimp Led to a Global Pandemic: The Rise of HIV," Youtube Video, 10:19, Nov. 29, 2017, https://www.youtube.com/watch?v=iz-womieBwG0.

Kaposi's sarcoma, with 26 men having been diagnosed with it in the past 2.5 years, some of which also had the fungal pneumonia, and 8 had died. At the time, scientists still didn't know what was making people sick or how it was spreading, but the association with gay men led to many referring to it as GRID – Gay-Related Immune Deficiency.[304] However, it wasn't limited to gay men, and people began to notice cases of it amongst hemophiliacs, heterosexual men, IV drug users, women, children, and 20 recent immigrants from Haiti, all of which hinted that the disease spread through blood.[305] By this point in 1982, the disease had been given the name AIDS,[306] and in March of 1983 the CDC issued a warning stating that the diseases was transmitted sexually through both gay and straight sex, and that doctors needed to be careful about blood transfusions.[307] By the summer of 1983, multiple groups had identified retroviruses in samples from AIDS patients, all of which were named differently as it wasn't clear they were the same virus; however, they were, and in 1986 the cause of AIDS received the name of HIV.[308] From there, people noticed that some other primates suffered from an AIDS like disease, and in 1985, after examining blood samples from Macaque monkeys, they found a virus similar to HIV that would later be named SIV, the Simian Immunodeficiency Virus, leading people to speculate that the virus we were now suffering from may have jumped the species barrier. Eventually it was determined that the virus had been transmitted to humans first around the turn of the 20th century through the handling of infected chimpanzee meat,[309] as the chimpanzee version of the virus was similar enough to have infected the hunter, leading it to be further passed on, providing it to mutate and evolve into the virus we're faced with.[310] Researchers believe that the main HIV infection present today came from infected chimpanzees that lived in Southwestern Cameroon, and estimate that HIV-1 first infected humans around 1908,

[303] Ibid.
[304] Ibid.
[305] Ibid.
[306] "Where did HIV come from? The AIDS Institute, https://www theaidsinstitute.org/education aids-101/where-did-hiv-come-0
[307] SciShow, "How a Sick Chimp Le to a Global Pandemic: The Rise o HIV," Youtube Video, 10:19, Nov 29, 2017, https://www.youtube com/watch?v=izwomieBwG0.
[308] Ibid.
[309] National Geographic, "AID 101," Youtube Video, 2:58, Nov 30, 2018, https://www.youtube com/watch?v=bjdqw9rXXd8.
[310] SciShow, "How a Sick Chimp Le to a Global Pandemic: The Rise o HIV," Youtube Video, 10:19, Nov 29, 2017, https://www.youtube com/watch?v=izwomieBwG0.

based on blood samples from the earliest known cases of HIV infection from 1959-1960.[311] From there, it's believed that in the 20s the virus was taken to the city of Kinshasa – called Leopoldville at the time – where it was further spread amongst the population through injectable drugs and prostitutes.[312] From there, it was only a matter of time before people left and spread the virus further, aided by the boom in air travel in the 60s,[313] and by the fact that it would still be another two decades before the virus was even recognized.

Though HIV is an awful virus that can lead to an early death, it has remarkably few symptoms beyond a brief period after initially becoming infected. When first infected, individuals may feel as though they're suffering from the flu, and may experience headaches, fevers, rashes, sore throats, and/or muscle and joint pain.[314] This is due to the fact that the virus is infecting a lot of cells, and the body is attempting to fight it off. That stage of infection is referred to as the Primary infection, and only lasts a few weeks at most before the symptoms pass.[315] After this point, people will feel fine for around a decade without any further sign of illness, because by then the person will have developed antibodies to keep the virus from running wild, all the while slowly losing CD4 T cells to the virus.[316] After this asymptomatic phase, also called the chronic stage of infection, is when people would be diagnosed with having AIDS, as once the CD4 count drops below a certain level, people suffer a greatly increased chance of opportunistic infections.[317] The reason HIV can progress like this overtime, slowly killing off a person's immune system without outward indication, is because of the way HIV evades the body's immune system. HIV makes itself a part of the DNA of immune system cells, and lays dormant, only replicating when the host cell is stimulated to react to an infection, making it so that these latently-infected cells aren't recognized by the immune

[311] Ibid.
[312] Ibid.
[313] National Geographic, "AIDS 101," Youtube Video, 2:58, Nov. 30, 2018, https://www.youtube.com/watch?v=bjdqw9rXXd8.
[314] SciShow, "How a Sick Chimp Led to a Global Pandemic: The Rise of HIV," Youtube Video, 10:19, Nov. 29, 2017, https://www.youtube.com/watch?v=izwomieBwG0.
[315] "The Science of HIV and AIDS - Overview," Avert, Oct. 10, 2019, https://www.avert.org/professionals/hiv-science/overview.
[316] SciShow, "How a Sick Chimp Led to a Global Pandemic: The Rise of HIV," Youtube Video, 10:19, Nov. 29, 2017, https://www.youtube.com/watch?v=izwomieBwG0.
[317] "The Science of HIV and AIDS - Overview," Avert, Oct. 10, 2019, https://www.avert.org/professionals/hiv-science/overview.

system as being infected, letting the virus exist as long as the cell lives, and allowing for the establishment of reservoirs of these cells in the lymph nodes, spleen, and gut.[318] Worse still, once established, HIV can continue to spread throughout these reservoirs without activating and triggering an immune response, infecting the immune system throughout the body, and passing it from cell to cell.[319] Thanks to these factors, HIV was able to spread for over half a century before being first noticed, and continues to persist despite the better part of 40 years of research into combating it.

Although HIV has spread across the world and infected people of all walks of life, not all groups of people have been impacted the same way by the virus. According to the World health Organization, the groups at greatest risk of infection are: homosexual men, people in closed settings like prison, transgender people, sex workers and their clients, and people who inject drugs, as well as young women and adolescent girls in some parts of Africa, and Indigenous peoples in communities.[320] In addition to people in these groups who have a greater risk of becoming infected, the WHO also noted that over two thirds of all people living with HIV live in Africa – particularly amongst the aforementioned groups in Africa – and that increased vulnerability is often linked to social and legal issues that cause people to be in high risk situations more often, preventing them from easily accessing treatment, prevention, and testing services.[321] Furthermore, because of the high rate of infection in Africa, the impact that has happened there has been far worse than for many other places around the world. The HIV/AIDS crisis has led to an "AIDS generation," where the majority of youth have been impacted significantly in some way by the virus, whether it be through knowing someone fighting it, knowing someone who died of it, or being orphaned by the virus.[322] Additionally, HIV has had further economic

[318] Ibid.
[319] Ibid.
[320] "HIV/AIDS," World Health Organization, Jul. 6, 2020, https://www.who.int/news-room/fact-sheets/detail/hiv-aids.
[321] Ibid.

and social impacts, damaging the African economy and terribly slowing efforts to fix it,[323] and creating issues of stigmatization and discrimination against both those infected within Africa, and against Africa as a whole.[324] When factoring in that many people living with HIV/AIDS in Africa can't access treatment,[325] and the disproportionate rates certain groups of people are affected by the virus, it's clear to see that, though it is a global concern and issue, there are certainly areas and groups hurt by this more than the rest of us.

We've come along away from when HIV was first identified nearly 40 years ago. Since then, we've developed numerous drugs and methods to treat HIV and prevent it from becoming AIDS, with our methods becoming more effective and with fewer side effects as we developed, even to the point where in a 2016 study, thanks to the antiretroviral therapy (ART) being employed, 1000 couples where one partner was HIV-positive were able to have unprotected sex without anyone becoming infected, due to the reduced levels of the virus in the infected partners thanks to ART.[326] However, despite these significant and noteworthy gains, there are still barriers to helping all those living with the virus, barriers which come unfortunately often in the form of human action. The responses to the HIV/AIDS pandemic from many serve as prime examples for the lows that humanity are capable of reaching, with some of the worst including Reverends and religious groups openly stating that the pandemic is punishment for allowing homosexual lifestyles and other 'sins,'[327] corporations taking the suffering of millions as an opportunity for profit,[328] individuals racializing the pandemic and blaming Africans for it the same way the black death was blamed on Jews,[329] and stories and cases of infected individuals who purposefully and vindictively spread the virus after becoming infected themselves.[330] Even simple examples of selfishness, like hiding your infection from your partner in order to have a baby or not being safe while being unfaithful, add to the

[322] Elias K. Bongmba, Facing a Pandemic: The African Church and the Crisis of AIDS (Waco, TX: Baylor University Press, 2007), 18-19.
[323] Bongmba, Facing a Pandemic, 19.
[324] Ibid., 11-12.
[325] Ibid., 142.
[326] SciShow, "Why HIV Isn't a Death Sentence Anymore," Youtube Video, 10:08, Dec. 13, 2017, https://www.youtube.com/watch?v=U52g6ZIRIW0
[327] Bongmba, Facing a Pandemic, 22.
[328] Ibid., 143.
[329] Ibid., 12.

spread of the virus and the damage it will cause. Thankfully though, as has been seen during pandemics and tragedies throughout history, for every Reverend who justifies the suffering there are others who reject that hateful and unchristian view and vocally advocate the need for treating HIV as the virus it is and not as a punishment, and who organize and pledge to fight against it.[331] For every corporation who chooses to exploit suffering and every bigot who chooses to blindly rejoice at the suffering of those different from them, there are those who do as much as they can to ensure others have access to the medications and treatments they need,[332] and there are Non-governmental, faith based, and international organizations working together to try and improve the lives of those suffering and ensure that as many people receive treatment as possible.[333] And for every individual who takes the power they have and uses it to harm others, you have individuals like Elias K. Bongmba, who dedicate years of their lives to researching and writing books meant to educate people on the suffering that is being experienced due to this pandemic, while simultaneously highlighting the factors contributing to the problem and providing suggestions and solutions to address them.

    With the capacity for hate and harm that we possess, it's easy to get lost in all of the awful things that happen and contribute to pandemics and the suffering of millions worldwide, as we're flooded with all of the different ways governments and leaders have failed their people, all the while further spreading the hate that only adds to the suffering. Navigating through all of that, it can be easy to lose sight of all the good that is being done and that can be done to combat it, because for every person who misguidedly or hatefully chooses to be selfish at the expense of others, there are others giving everything they have to combat the suffering, and the many injustices that contribute to it.

[330] Ibid., 12-13, 83-84.
[331] Ibid., 22-23.
[332] Ibid., 143-144.
[333] Ibid., 163-164.

XIII.
HIV/AIDS AND CHRISTIANITY

> **"** Discrimination, moralism, rejection, and mystification are ethically the most serious threats when encountering the [HIV] epidemic."[334]

[334] Paula Clifford, Theology and HIV/AIDS Epidemic, Christian Aid, 2004, p. 9.

[335] This form of theology aims to provide a framework that shapes the proper Christian response to HIV/AIDS. A theology of HIV/AIDS often attempts to deal with the issues of discrimination and stigmatizaiton that are commonly experienced by individuals who are HIV positive.

[336] Van Wyngaard, Arnau. "Towards a Theology of HIV/AIDS". Verbum et Ecclesia. 27. 2006, 10.4102/ve.v27i1.148.

The HIV/AIDS epidemic is a devastating entity that has forced Christians to reconcile their divine mission within the secular world. This crisis has a distinct theological dimension that calls the religious faithful to be a positive presence in the lives of those negatively impacted by disease and suffering. Thus, there is a growing call in Christian circles for what is commonly being referred to as a "theology of HIV/AIDS."[335] In this chapter, I will focus on some of the prominent theological issues facing church bodies with the advent of HIV/AIDS, and how Christian communities have responded to this epidemic in various parts of the world, particularly in Africa, Western Europe, and North America.

It is difficult to speak of Africa without discussing how the HIV/AIDS epidemic has influenced its people. Pastor of the Swaziland Reformed Church, Arnau van Wyngaard, suggests the growing need for a properly formulated theology for this epidemic as Christians can no longer speak about the church without discussing the effects of HIV/AIDS.[336] Behind the devastating statistics describing the extent of the HIV/AIDS epidemic lie concrete situations of human pain, suffering, and death. These distressing circumstances have generated a multitude of theological questions, reigniting debates about the existence of evil and suffering in the world with the existence of a loving and personal God. Van Wyngaard argues for the necessity of the church's response to epidemics, not only practically, but theologically, as well, since a:

> "[F]AILURE TO PROBE THE THEOLOGICAL SIGNIFICANCE OF THIS MOMENT WILL BE NOT ONLY A MISSED OPPORTUNITY BUT ALSO IRRESPONSIBLE. JUST AS THE ENTIRE CHRISTIAN WORLD HAS BEEN AND CONTINUES TO BE MOBILIZED IN PROGRAMMES MEANT TO COMBAT RACISM, SEXISM, AND ECONOMIC EXPLOITATION AND CULTURAL ARROGANCE, WE NOW NEED THEOLOGIES THAT WILL HELP US DEAL WITH THE CHALLENGE OF HIV/AIDS."[337]

There was a need for a theological response to the challenges that PWA, people with AIDS, faced, particularly with the growing awareness of the disease's "intensely spiritual dimension."[338] Thus, responses were not just for medical intervention, but also for pastoral and spiritual care, as well. Although enduring condemnatory and critical attitudes still exist within the majority of Christian groups, recent field research suggests that church bodies have transformed into natural referral centres for PWA, particularly in Tanzania, Uganda, and Kenya.[339] Responding to the Christian vocation to love, these church bodies imitated Jesus when He welcomed the sick and disabled with open arms.[340] By focusing on Jesus' loving response and compassion towards the ill and suffering, these Christians have been able to appropriately respond to their vocation to love, actualizing the very identity and mission of the Church. Attentive to the theological deficiencies of the Church's initial response to PWA, Ugandan theologian, John Mary Waliggo, emphasizes the importance of counselling, visiting and showing solidarity with the sick, caring

[337] Maluleke, T S s.a.. Towards an HIV/AIDS-Sensitive Curriculum. <http://www.wcc-coe.org/wcc/what/mission/dube-7.html>

[338] "From Crisis to Kairos: The Mission of the Church in the Time of HIV/AIDS, Refugees and Poverty", Mission Studies 25, 2: 311-312. p. 129.

[339] Shorter, Aylward, and Edwin Onyancha. The Church and AIDS in Africa: A Case Study. Nairobi City. Nairobi, Kenya: Paulines Publications Africa, 1998. p. 79.
Matthew 25:31-46.

[340] "From Crisis to Kairos: The Mission of the Church in the Time of HIV/AIDS, Refugees and Poverty". p. 130.

for orphans, providing liberative and respectful assistance, and Christian preaching on matters such as contraception and sex.³⁴¹ For an effective HIV/AIDS ministry within the Church, there needs to not only be a substantial amount of financial and material support, but also a great deal of personal support, i.e., time, energy, and commitment, as "each member of the community and disciple of Christ carries out his or her unique function in the name of the church."³⁴² With the advent of HIV/AIDS, it has become increasingly apparent that the epidemic disproportionately targets women as they often "carry the greater burden of care."³⁴³ For instance, there are more women that end up taking care of PWA and AIDS orphans. There are also a higher number of women involved in both organized and non-organized HIV prevention programmes that emphasize care for PWA.³⁴⁴ British Roman Catholic priest, Rev. Kevin Kelly, states that with the impact of HIV/AIDS on women and their role in society, churches must strive to adopt a more "feminine ecclesial identity" that calls for a greater emphasis on the experiential accounts of women as they have historically been neglected in discussions concerning church policy.³⁴⁵

³⁴¹ "From Crisis to Kairos: The Mission of the Church in the Time of HIV/AIDS, Refugees and Poverty". p. 130.
³⁴² Ibid., 131.
³⁴³ Ibid., 134.
³⁴⁴ Ibid.
³⁴⁵ Ibid., 136.
³⁴⁶ Ibid.
³⁴⁷ Ibid., 137.

" LIKE MANY WOMEN WHO HAVE COMMITTED THEM-SELVES TO THE CAUSE OF FACING THE CHALLENGES POSED BY THE EPIDEMIC, THE CHURCH DISCERNS A CALL TO BE IN THE MIDST OF GOD'S PEOPLE WHO ARE TRAPPED IN THE THROES OF SICKNESS, SUFFERING AND DEATH. AS MANY PWA TESTIFY, WHAT RENEWS THEIR HOPE IN LIFE IS THE FACT THAT SOME PEOPLE CARE ENOUGH TO VISIT THEM, TO BE WITH THEM. BY THEIR SIMPLE BUT COURAGEOUS PRESENCE, WOMEN EMBODY THE FACE OF THE CHURCH AS A LOVING, CARING PARENT WHO NEITHER REJECTS NOR ABANDONS HER OWN (CF. IS 49:15)."³⁴⁶

Therefore, with the HIV/AIDS epidemic there now

exists a greater movement towards a more feminine or pro-women theology that emphasizes the significance of women in the fields of pastoral, theological, and sacramental care.[347] This movement has often included the integration of women in various sacramental ministries, i.e., reconciliation, anointing of the sick, and communion.[348] This emphasis on pastoral and theological femininity during the HIV/AIDS epidemic serves to demonstrate the many lay and religious women that have responded to the Christian vocation to love in their communities. Thus, it can be said that the theology of HIV/AIDS would be lacking if it did not emphasize the role of women in the church and in the spreading of Christ's gospel.

PWA have often requested the sacrament of reconciliation as a means of spiritual, emotional, and physical healing. In the context of HIV/AIDS, the sacrament of reconciliation is typically used to forgive any past actions or behaviors that may have led to a person being infected; help with the reconciliation of families that have been affected; and ultimately, cultivate a reconciliation with God, especially in view of a person's impending death.[349] The arrival of HIV/AIDS brought with it a profound sense of guilt, shame, and fear that could not be alleviated with mere medical or psychological intervention; thus, the sacrament of reconciliation has been given as a way to provide the spiritual assistance and comfort that many PWA need. As Rev. Edward Philips put it:

> AIDS IS AN ILLNESS, WHICH IMPACTS ALL PARTS OF THE PATIENT'S LIFE... MANY OF OUR PATIENTS FEEL THE FEAR AND PAIN OF ALIENATION FROM SELF, FAMILY, COMMUNITY AND GOD. IT IS THE UNDERLINING COMPONENT OF OUR WORK THAT WE OFFER AND REFLECT UNCONDITIONAL LOVE AND ACCEPTANCE OF OUR PATIENTS. THIS REFLECTS BOTH A THEOLOGICAL AND PSYCHO-SOCIAL APPROACH TO THE PATIENTS. IT IS OUR BELIEF AND PRAXIS THAT THE UNCONDITIONAL LOVE OF GOD AND CHRIST FOR OUR PATIENTS CAN ONLY BE

[348] Ibid.
[349] Ibid., 141.
[350] Ibid.

EXPERIENCED THROUGH OUR UNCONDITIONAL LOVE AND ACCEPTANCE OF THEM. THIS PROCESS LEADS TO RECONCILIATION AND HEALING OF SELF, FAMILY, COMMUNITY AND GOD"[350]

To appropriately respond with love, it is important that individuals are aware of the various circumstances that PWA face due to absolute or situational poverty.[351] Individuals experiencing absolute or situational poverty often experience emotional and social challenges, acute and chronic stressors, cognitive lags, and various health and safety issues.[352] These factors often lead to poverty playing a significant role in the increased stigma and transmission of HIV/AIDS, and vice versa.[353] As Eric Jensen writes in *Teaching with Poverty in Mind*, "[a] problem created by poverty begets another, which in turn contributes to another, leading to a seemingly endless cascade of deleterious consequences."[354] Poverty appears to both increase and affect the occurrence of HIV/AIDS in two ways: firstly by producing situations where more people get infected and secondly, by delaying the process of treatment for HIV/AIDS.[355] Often isolated from their families or the stabilizing influence of their communities, individuals living in poverty become susceptible to having casual sex with PWA, or risk spreading the disease if they themselves have it. It is also the case that when faced with poverty, many individuals turn to prostitution as a means to alleviate their suffering. For many years, the Church has warned against the evils of sexual immorality, i.e., pre- and extra-marital sexual relations; however, warnings like these have often generated a significant amount of fear in PWA. As a result of this fear, many PWA are hesitant to seek help for their illness because of the stigma that is attached to it.

In the West, HIV/AIDS was simply viewed as "a gay plague" as it was first encountered primarily in homosexual communities.[356] The ignorance from this belief left

---

[351] "Absolute poverty" refers to a scarcity of such necessities as shelter, food, and water, whereas "situational poverty" refers to a type of poverty that is caused by sudden crisis or loss. Although situational poverty is often temporary, it can lead to absolute poverty.

[352] Jensen, Eric. Teaching with Poverty in Mind: What Being Poor Does to Kids' Brains and What Schools Can Do About It. , 2009. p. 7.

[353] When I use the term "poverty", I am also referring to the illiteracy and lack of education that is often a byproduct of absolute poverty.

[354] Jensen, Eric. Teaching with Poverty in Mind: What Being Poor Does to Kids' Brains and What Schools Can Do About It. , 2009. p. 7.

[355] Van Wyngaard, Arnau. "Towards a Theology of HIV/AIDS" (2006).

[356] Paula Clifford, Theology and HIV/AIDS Epidemic, Christian Aid, 2004, p. 10.

many communities vulnerable to infection, and caused many PWA to avoid discussing their HIV status for fear of being rejected by their religious and social communities.

As Paula Clifford writes in *Theology and HIV/AIDS Epidemic*:

> "CHRISTIANS IN THE 1980S WERE CONFRONTED WITH A NEW TEACHING IN MANY CHURCHES: HIV IS GOD'S PUNISHMENT FOR SEXUAL SINNERS... [THIS TEACHING] WAS ADOPTED WITH ENTHUSIASM, BOTH IN EUROPE AND NORTH AMERICA, PARTICULARLY — BUT NOT EXCLUSIVELY — BY EVANGELICAL GROUPS. IT ALSO GAINED GROUND IN SOME DEVELOPING COUNTRIES, WHERE PEOPLE TYPICALLY SEEK AN EXPLANATION (SUCH AS BAD SPIRITS) FOR ANY KIND OF MISFORTUNE."[357]

For many Christians, this stigmatization towards sexual immorality led to silence being the preferred option for many PWA as, all too often, the alternative was public and societal rejection. While most reasonable theologians would not support such an extreme viewpoint which links HIV/AIDS directly to God's judgment, there are many individuals in the Church who believe that individuals pursuing an immoral lifestyle, i.e., homosexuality or the use of intravenous drugs, are getting their "just reward" by being HIV positive.[358] This has led many to believe that the Church is largely to blame for the stigma and the spread of HIV/AIDS since its theology on human sexuality is closely intertwined with the concepts of sin, guilt, and punishment.[359] Although this may be the case, we must not forget the official responses of the Church to the HIV/AIDS epidemic.

In 1989, Pope John Paul II made frequent appeals to avoid discriminatory treatment of PWA. During his address given at Mision Dolores, he proposed the unconditional love of God Himself as a model for the Christian response to those suffering from HIV/AIDS. He stated that:

[357] Ibid., 10-11.
[358] Van Wyngaard, Arnau. "Towards a Theology of HIV/AIDS" (2006).
[359] Paula Clifford, Theology and HIV/AIDS Epidemic, Christian Aid, 2004, p. 11.

"GOD LOVES YOU ALL, WITHOUT DISTINCTION, [AND] WITHOUT LIMIT ... HE LOVES THOSE OF YOU WHO ARE SICK, THOSE SUFFERING FROM AIDS. HE LOVES THE FRIENDS AND RELATIVES OF THE SICK AND THOSE WHO CARE FOR THEM. HE LOVES ALL WITH AN UNCONDITIONAL AND EVERLASTING LOVE."[360]

A year later, he continues on to say that:

"AIDS THREATENS NOT JUST SOME NATIONS OR SOCIETIES BUT THE WHOLE OF HUMANITY. IT KNOWS NO FRONTIERS OF GEOGRAPHY, RACE, AGE, OR SOCIAL CONDITION ... THE THREAT IS SO GREAT THAT INDIFFERENCE ON THE PART OF PUBLIC AUTHORITIES, CONDEMNATORY OR DISCRIMINATORY PRACTICES TOWARD THOSE AFFECTED BY THE VIRUS OR SELF-INTERESTED RIVALRIES IN THE SEARCH FOR A MEDICAL ANSWER, SHOULD BE CONSIDERED FORMS OF COLLABORATION IN THIS TERRIBLE EVIL WHICH HAS COME FROM HUMANITY."[361]

In 2005, Pope Benedict XVI's response to the HIV/AIDS epidemic emphasized the need for further education and training in the field of HIV/AIDS as: "[p]rogress becomes true progress only if it serves the human person and if the human grows, not in terms of technical power alone, but in moral capacity too."[362] Following in the footsteps of Benedict XVI, Kenyan Bishops have stated their objectives as follows:

"THE CHURCH IN KENYA WILL STRIVE TO INCREASE GENERAL CAPACITY FOR HIV PREVENTION THROUGH FORMATION FOR RESPONSIBLE AND RESPECTFUL BEHAVIOR AS WELL [AS] EDUCATION ABOUT HIV, AIDS AND SEXUALITY... IN ORDER TO DIMINISH INEQUALITIES AND ABUSE WHICH FUEL THE SPREAD OF HIV, THE CHURCH WILL PROMOTE EDUCATION OF WOMEN, THEIR EMPOWERMENT AND LEADERSHIP. CONCOMI-

[360] Pope John Paul II, Address given at Mision Dolores, 1989.
[361] Pope John Paul II, visit to Tanzania, September 1990.
[362] Carey, Timothy James. Muslim and Catholic Responses to HIV and AIDS in Kenya. p. 7.

TANTLY, THE CHURCH WILL IMPART LIFE SKILLS TO MEN, WOMEN, CHILDREN, AND YOUTH, ENABLING THEM TO MAKE INFORMED AND RESPONSIBLE DECISIONS."

In Kenya, a significant proportion of education, healthcare, and medical treatment for PWA is offered by various Catholic clinics, hospitals, parishes, and other institutions.[363] This visible presence of Catholics amidst the HIV/AIDS crisis affirms the dual entity of the Church as both a religious and a social institution that aims to attend to the bodily, social, and spiritual dimensions of the human condition. When the Church speaks of education in the context of HIV/AIDS and sexuality, it is often associated with stigma reduction as there exists a significant amount of discrimination in the religious and secular world directed towards PWA. This is often due to a line of thinking that involves a moralistic interpretation of the situation that seeks to blame PWA for contracting the disease.[364] The Church responds to this by reinforcing the belief that "[n]o one on earth has access to another's conscience, and no one can judge whether or not an individual has sinned"; therefore, individuals must strive to avoid these divisive and hurtful ways of thinking about those who are or have been inflicted with HIV/AIDS.[365] For bishops around the world, stigma reduction is at the forefront of the Church's response to the epidemic as the vocation to love demands Christians "[t]o include the excluded… and to embrace and touch the stigmatized."[366] According to Catholic social teaching, followers of Christ must always hold to the inherent dignity of the human person as this can never be taken away regardless of how many sins have been committed by the individual.[367] As Timothy Carey puts it: "Stigma and discrimination are thus *anathema*[368] to Catholic social teaching."[369]

There is an inherent level of alienation that occurs when discussing events of centuries past. The numbers and

[363] Carey, Timothy James. Muslim and Catholic Responses to HIV and AIDS in Kenya. Lanham, MD: Lexington Books, 2018. p. 7.
[364] Ibid., 9.
[365] Ibid., 9-10.
[366] Ibid., 10.
[367] The principle of human dignity is considered to be the foundational principle of Catholic social teaching.
[368] "Anathema" is a term commonly used to indicate that a position or teaching is in contradiction to Catholic faith and doctrine.
[369] Ibid.

history of the Black Death or Spanish Flu can be described in intricate, layered detail but will ultimately fail to accurately capture what it was like to live through that time. Even then, is a person's first-hand experience too limited in scope to capture the breadth of an event as massive as a global pandemic? What if a person has access to a constant stream of up-to-date pandemic updates and news from around the world? Are they now able to understand the entirety of the situation? With COVID-19, anyone with a phone can easily track how governments and individuals are reacting to a global pandemic in real time. Gone are the days of slow, localized sources of information; now almost anyone can instantly pull up pandemic updates, government regulations, opinion pieces, and presidential tweets about the pandemic anytime they want. What kind of an effect does this individual informational empowerment have on how a global pandemic plays out? How we, as a people, react to this crisis has consequences in both the short term with regards to keeping people safe right now, and in the long term, shaping how we respond to future crises. We will discuss in this chapter the symptoms, methods of transmission, prevention, and treatment of COVID-19, as well as some of the notable societal and cultural reactions to the pandemic, both positive and negative.

Coronavirus disease 2019, or COVID-19 for short, is the official name for the disease caused by the virus designated as severe acute respiratory syndrome coronavirus 2 (SARS-CoV-2). As is apparent, SARS-CoV-2 is part of the coronavirus family of viruses, which are characterized by their roughly spherical shapes and "coronas" of viral spikes. Coronaviruses are typically endemic among bat, rodent, or bird populations, from which they sometimes spread to larger mammals, including humans. While symptoms vary among different species, in humans coronaviruses typically cause respiratory tract infections. Outside of COVID-19, the most significant epidemics caused by coronaviruses have been the 2012 outbreak of Middle East respiratory syndrome (MERS) caused by MERS-CoV and the 2002-2004 outbreak of severe acute respiratory syndrome (SARS) caused by SARS-CoV-1. Like SARS, COVID-19 originated in China, with the point of origin of the latter being the city of Wuhan in Hubei province. From Wuhan, the virus spread throughout China and eventually, the rest of the world. At the start of the year, on January 23 2020, there were 571 cases in China and 10 in other countries. Six months later, on June 28, the total global number of confirmed COVID-19 cases crossed 10 million. By comparison, there were over 8000 confirmed cases of SARS between 2002-2004, and 2519 cases of MERS since 2012. Coronaviruses vary greatly in terms of lethality and transmissibility, with milder variants causing common colds while more dangerous coronaviruses cause SARS, MERS, COVID-19, among others. COVID-19 has a significantly lower fatality rate than SARS or MERS but has a higher transmission rate, as well as a period in which a newly infected person may be infectious without showing symptoms. It is these factors which have led to COVID-19 spreading so rapidly around the world.

One barrier in the way of diagnosing COVID-19 is that it shares many symptoms with other common illnesses, such as the common cold or seasonal flu. The most common symptoms of COVID-19 have been determined to be: fever, coughing, and fatigue. Less-common symptoms include nausea, headaches, bodily aches and pains, loss of taste and/or smell, and vomiting. As is typical of many infectious diseases, there is a period of incubation between when a person is initially infected and when they begin to display symptoms. For COVID-19, this period typically lasts around a week, though it can be short as a day or as long as two weeks. A minority of infected cases have been asymptomatic, i.e. show no symptoms at all. The exact fatality rate of COVID-19 is unknown, due to both how recently it was discovered and the fact that many cases go undiagnosed. Early estimates place the case fatality rate of COVID-19 at around 1-2%.[370] The disease has been found to have a significantly higher fatality rate in individuals over 60 years of age and those who have underlying conditions such as hypertension, diabetes, Asthma, or serious disability. Severe cases of COVID-19 may require the application of a ventilator. COVID-19 is primarily spread via airborne droplets expelled by coughing, sneezing, or talking, which are then inhaled or ingested by people within close proximity. Less commonly, a person can become infected via contact with a droplet-contaminated surface, especially if they touch their eyes, nose, or mouth without having washed their hands.[371] As there is no cure or vaccine available at present, international efforts to combat the virus have focused on policies of "social distancing" in order to prevent opportunities for the virus to be transmitted from person-to-person. Countries have mandated varying levels of social distancing, often changing from moment-to-moment as case numbers in the country rise or fall. The most basic and effective method of social distancing is complete social isolation, i.e. not coming into contact with anyone who doesn't already live in your home. Other methods of social distancing include: wearing a mask to avoid spreading or inhaling expelled droplets, maintaining at least 6 ft. distance from other people in public so as to stay out of the range in which droplets can be spread, the shutting down of non-essential workplaces, avoiding public transportation, avoiding travel outside of the city or town where you live, and so on. On paper, the goal of social distancing is to completely prevent any and all chances for the virus to spread. Realistically, stopping all spread is impossible, and instead the goal becomes to "flatten the curve" as much as possible. Flattening the curve means to slow the spread of

---

[370] Lionel Roques et al. "Using Early Data to Estimate the Actual Infection Fatality Ratio from COVID-19 in France," Biology 9, no. 5 (2020): 97, https://doi.org/10.3390/biology9050097.

[371] "An introduction to SARS-CoV-2," NCCEH, last updated May 29, 2020, https://ncceh.ca/documents/evidence-review/introduction-sars-cov-2.

[272] Elisabeth Rosenthal, and Lawrence Altman, "China Raises Tally of Cases and Deaths in Mystery Illness," The New York Times, March 27, 2003, https://www.nytimes.com/2003/03/27/world/china-raises-tally-of-cases-and-deaths-in-mystery-illness.html.

[373] Frank Ching, "Sars Gives China a Lesson in Globalization," YaleGlobal Online, May 2, 2003, https://yaleglobal.yale.edu/content/sars-gives-china-les-

the virus such that the number of infected people at any given moment never exceeds the capability of the country's health care system. If social distancing is not mandated or practiced, the number of infected people may quickly peak, overwhelming the health care system. In an overwhelmed health care system, great risk is faced both by people infected with COVID-19 and people requiring medical treatment for unrelated conditions. The introduction of social distancing around the globe has not been without controversy. Effective social distancing requires action on both the government's part (issuing of stay-at-home orders, closing of businesses, offering benefits for the unemployed, and so on) and action on the part of ordinary people to distance themselves socially. The economic effect of social distancing is not to be understated either, as closing businesses is unsurprisingly not good for the economy and puts many people out of jobs. Rather than generalize, we will now look at several notable reactions to the pandemic by countries, and their successes or failures to combat COVID-19.

When SARS broke out in China in November 2002, the government did its best to cover up the incident and prevent knowledge of the outbreak from spreading outside of the originally affected province.[372] Following this, the Chinese government promised that it would not repeat the mistakes it had made in handling the SARS outbreak.[373] With COVID-19, China somewhat upheld this promise, though the nation was not free from international criticism. The government managed to accrue significant criticism before the virus had even been identified. On December 30 2019, Dr. Li Wenliang sent a message in a group chat warning fellow doctors about a virus he encountered which appeared similar to SARS. In response, Chinese police warned him to stop "making false comments" and that he had "severely disturbed the social order."[374] Criticism of the Chinese government in and outside of the coun-

son-globalization.

[374] "Li Wenliang: Coronavirus kills Chinese whistleblower doctor," BBC News, February 7, 2020, https://www.bbc.com/news/world-asia-china-51403795.

[375] Li Yuan, "Coronavirus Crisis Exposes Cracks in China's Facade of Unity," The New York Times, Jan 31, 2020, https://www.nytimes.com/2020/01/28/business/china-coronavirus-communist-party.html.

[376] Charlie Campbell, and Amy Gunia, "China Says It's Beating Coronavirus. But can We Believe Its Numbers?," TIME, April 1, 2020, https://time.com/5813628/china-coronavirus-statistics-wuhan/.

[377] Niall McCarthy, "COVID-19 Deaths Per 100,00 Inhabitants: A Comparison," statista, June 23, 2020, https://www.statista.com/chart/21170/coronavirus-death-rate-worldwide/.

[378] Steve Almasy, "Another day of record coronavirus cases as more states rethink mask mandate," CNN Health, 3 July, 2020, https://www.cnn.com/2020/07/02/health/

try only intensified when Dr. Li died after being infected with COVID-19. Criticism was also levied at what the Chinese populace saw as a disorganized and random initial response to the outbreak.[375] Later, the government's rigorous handling of the crisis would be praised, as well as the improvement in open communication between Chinese and international scientists. As of June 2020, COVID-19 has been largely eradicated in China, with among the lowest total case numbers in the world. However, given the Chinese government's propensity for bending published statistics and news to suit their needs, many understandably have shown distrust towards Chinese case numbers, believing that they should be higher than they currently are stated to be.[376] As with most of the countries currently facing COVID-19, it is difficult to exactly parse at the moment how effective a government's response is, which is only magnified by the narrative-twisting nature of the Chinese government. Regardless, going off of current information it appears as though China's response to COVID-19 is far from unsuccessful, despite the significant amounts of (often-valid) criticism and controversy.

The most significant trait seen in the response of the United States to COVID-19 is that of denying certain realities for the sake of economic growth or preserving individual liberties. The U.S. currently has the highest number of cases in the world and one of the highest case-fatality ratios.[377] Many U.S. states pushed forward with reopening plans despite high numbers of new infections.[378] Anti-lockdown protests, some with thousands of attendees, have been organized across the country. These protests have largely been organized by conservative groups and have been endorsed by President Trump on more than one occasion.[379,380] Objectives of the protests range from the prevention of an economic collapse or the loss of small business to opposition towards state governments "infringing on individual liberties" with lockdown orders.[381] Connections

us-coronavirus-thursday/index.html.

[379] Michael Martina et al., "How Trump allies have organized and promoted anti-lockdown protests," Reuters, April 21, 2020, https://www.reuters.com/article/us-health-coronavirus-trump-protests-idUSKCN2233ES.

[380] Phil Thomas, "'Great people!' Trump backs anti-lockdown protesters filmed harassing reporter in Long Island" The Independent, May 16, 2020, https://www.independent.co.uk/news/world/americas/us-politics/trump-lockdown-protest-twitter-video-reporter-journalist-a9518296.html.

[381] Russell Berman, "What the 'Liberate' Protests Really Mean for Republicans," The Atlantic, April 23, 2020, https://www.theatlantic.com/politics/archive/2020/04/coronavirus-protests/610363/.

[382] Joseph Uscinski and Adam Enders, The Coronavirus

have been drawn between these protests and COVID-19 conspiracy theories promoted by members of the far-right. Theories include:the idea that the virus was created and released by China on purpose, that the pandemic is a hoax perpetrated by the powers that be, and most outlandishly that the disease is not caused by a virus but is in fact a reaction caused by new 5G cell phone towers.[382] Theories relating to the Chinese origins of the virus have further aggravated anti-Chinese sentiments in the U.S., contributing to a wave of harassment and racism towards Chinese people and other Asians seen worldwide due to the virus. From the perspective of non-Americans, it can be easy to look at the situation in the U.S. as an amusing confirmation of the "stupid Americans" stereotype. The sad reality is, however, that in a pandemic this kind of self-centered denial of scientific realities has deadly consequences. Anytime anybody in any country needlessly breaks quarantine or fails to follow social distancing procedures, even if they themselves are healthy and not at significant risk, they are effectively gambling with the lives of the more vulnerable members of society. What one should take away from the U.S.'s response to COVID-19 is not some humorous affirmation of American stereotypes but a warning about the dangers of prioritizing individual freedoms over the safety of the vulnerable, and how important it is to elect governments that won't reinforce such reckless behaviour.

Following the theme of posing great danger to the vulnerable members of society, the COVID-19 pandemic has posed a significant threat in developing countries around the world. Imagery of the Black Death is evoked from the situation in Ecuador in April, where hospitals, morgues, funeral homes, and cemeteries were so overwhelmed that dead bodies were literally abandoned in the street.[383] The situation looks extremely bleak in Brazil with case numbers and a death toll both second only to the U.S.'s, and a volatile

Conspiracy Boom, The Atlantic, April 30, 2020, https://www.theatlantic.com/health/archive/2020/04/what-can-coronavirus-tell-us-about-conspiracy-theories/610894/.

far-right president who actively pushed against quarantine efforts and denied that COVID-19 posed any significant danger.[384] The international response to COVID-19 is only as strong as its weakest links, and unfortunately such links have been highlighted due to COVID-19. In terms of more effective responses to COVID-19, South Korea demonstrated an extremely strong and immediate response very early on in the pandemic. The country used intrusive contact tracing, extensive testing, and strict isolation procedures to help eradicate the virus.[385] Germany achieved similar success with less intrusive contact tracing and a phased-in lockdown, eventually doing so well that Germany was able to accept patients from Italy and France.[386] New Zealand was declared virus-free for a period in early June, allowing social distancing policies to be lifted. That being said, at time of writing, South Korea, Germany, and New Zealand have each experienced second waves of varying severities.[387] Preventing a disaster is unglamourous. It doesn't make the news when another new bridge is intelligently and safely built, but it does when a bridge collapses and kills ten people due to shoddy planning. Minimizing an outbreak of COVID-19 in a region using robust enforcement of social distancing and individual citizens doing their part doesn't usually make for a good headline, but a massive outbreak caused by a few people deciding to break quarantine does. The reason why this is is probably mostly because we tend to pay more attention to bad news than good news, and news outlets wanting to retain viewership make use of that fact.[388] It's possible that this kind of negative news could help highlight teachable moments for people; showing causes of outbreaks and what could have been to avert them. However, another more dangerous outcome of this type of reporting is a shift towards solving something after it's become a problem, instead of ensuring that the problem never occurs in the first place. Reporting on outbreak after outbreak begins to imply that these

[384] Andrew Nikiforuk, "Brazil's Descent into COVID-19 Hell," The Tyee, June 10, 2020, https://thetyee.ca/News/2020/06/10/Brazil-Descent-COVID-Hell/.

[385] Derek Thompson, "What's Behind South Korea's COVID-19 Exceptionalism," The Atlantic, May 6, 2020, https://www.theatlantic.com/ideas/archive/2020/05/whats-south-koreas-secret/611215/.

[386] Margaret Evans, "Germany, a global leader on COVID-19 response, cautiously comes out of lockdown," CBC, May 30, 2020, https://www.cbc.ca/news/world/germany-coronavirus-covid-lockdown-merkel-1.5590731.

[387] Nikiforuk, "Brazil's Descent."

[388] Stuart Soroka et al., "Cross-national evidence of a negativity bias in psychophysiological reactions to news," PNAS 116, no. 38 (September 2019): 18888-18892, https://doi.org/10.1073/pnas.1908369116.

events are inevitable, and that focus should be directed towards finding a cure or vaccine. The problem with this is that it is difficult to tell when an effective vaccine will be able to be tested, manufactured, and distributed, and in that in the meantime the virus will still be spreading and killing people. That's not to say that a vaccine shouldn't be developed, but simply that preventative measures, while less glamorous, ultimately lower the need for an effective vaccine while still saving countless lives. It's a safe bet that future global crises will likely also disproportionately affect the most vulnerable members of the world's population, thus for their sake as well as everybody else's, we must collectively work to prevent or lessen as many disasters as possible.

XV.
COVID-19 AND CHRISTIANITY

> **WE FIND OURSELVES AFRAID AND LOST. LIKE THE DISCIPLES IN THE GOSPEL WE WERE CAUGHT OFF GUARD BY AN UNEXPECTED, TURBULENT STORM. WE HAVE REALIZED THAT WE ARE ON THE SAME BOAT, ALL OF US FRAGILE AND DISORIENTED, BUT AT THE SAME TIME IMPORTANT AND NEEDED, ALL OF US CALLED TO ROW TOGETHER, EACH OF US IN NEED OF COMFORTING THE OTHER. ON THIS BOAT . . . ARE ALL OF US."**
> *—POPE FRANCIS, URBI ET ORBI MESSAGE, MARCH 27, 2020*[389]

[389] Pope Francis "Urbi et Orbi Message", March 27, 2020, http://w2.vatican.va/content/francesco/en/messages/urbi/documents/papa-francesco_20200327_urbi-et-orbi-epidemia.html

As mentioned in the previous chapter, there is something unique about living in a time of pandemic. Looking back on history it can be easy to judge people's actions, or lack thereof, but if we stop to reflect for a moment, that is exactly what people will be doing when they look back on right now. What we call the present is soon to become the past. In this chapter we'll be looking at some of the effects COVID-19 has had on the life of the Church as we speak. How has the Catholic Church in particular responded on an international scale; what impact has the coronavirus had on public worship, both now and going forward; and what, exactly, are Christians called to do given the current situation? Unlike many strictly secular entities such as governments and public health organizations, the Church must not only be attentive to the physical, mental, and emotional aspects of people's well being, but it is also responsible for them spiritually. If nothing else, the current situation has certainly demonstrated how truly delicate of a balancing act that can be.

To begin with, it is worth noting that Easter and Christmas are usually the only two times each year when it is normal for churches to be full to capacity. The rest of the year attendance tends to fluctuate, however, in much of Euprope and North America it has been gradually decreasing for a number of years now, a trend being watched quite closely by Catholic sociologist Dr. Stephen Bullivant.[390] That being said, for the first time in living memory, in many places, including St. Peter's at the Vatican—which usually draws a crowd of anywhere from 15,000 to 80,000 people—Easter was celebrated in practically empty churches. This was not just because there was a particularly small congregation this year—there simply was no congregation. Catholic bishops around the world gradually began suspending public celebrations of the Mass, in some places as early as late February. By the time Holy Week rolled around in April, many countries around the world had issued self-isolation orders of some sort, putting into effect restrictions on public gatherings unlike anything seen since the Spanish Flu pandemic of 1918. For the most part, there was a fairly unified consensus among the bishops—a rare occasion in itself—and they chose to suspend the ordinary obligation to attend Sunday Mass, which, at time of writing, is still in effect in many places. This decision, though not taken lightly, turned out to be fairly controversial. Many people considered it necessary in order to help "flatten the curve" and avoid spreading the virus. If the Catholic Church claims to be the greatest advocate for the value of human life in the world today, then it is imperative that it not contribute to the spread of the virus by holding public gath-

[390] See Bullivant, Catholicism in the time of Coronavirus, Chapter 2.

erings. Others argued that by permitting the state to determine which services were deemed 'necessary'—without contesting the fact that in Italy, for example, tobacco shops were permitted to remain open while churches were forced to close—it was effectively sending the message that the Church is not in-fact an essential service, and that religion is simply a personal matter with nothing to offer society as a whole. In either case, however, by discouraging people from going against the guidelines set by public health officials, the Catholic Church sent a very strong message to the world that the Church was prepared to listen to the advice of the medical community. Which could be said to indicate how, contrary to popular belief, it does not consider religion and science to be opposed to one another.[391] It is difficult to grasp why suspending public Masses is such a big deal if one does not understand what Catholics actually believe about the Mass. When Gov. Charles Henderson suspended public gatherings in the state of Alabama due to the influenza pandemic of 1918, Fr. James E. Coyle addressed a letter to the Catholics of the diocese of Birmingham who were deprived of the opportunity to attend Mass. In it he says:

> "I TRUST FROM THIS VERY CIRCUMSTANCE, (YOU WILL) APPRECIATE MORE THOROUGHLY WHAT HOLY MASS IS FOR THE CATHOLICS. SUNDAY SERVICE IS NO MERE GATHERING FOR PRAYER, NO COMING TO A TEMPLE TO JOIN IN HYMNS OF PRAISE TO THE MAKER, OR TO LISTEN TO THE WORDS OF A SPIRITUAL GUIDE, POINTING OUT HE MEANS WHEREBY MEN MAY WALK IN RIGHTEOUSNESS AND GO FORWARD ON THE NARROW WAY THAT LEADS TO LIFE ETERNAL. NO, THERE IS SOMETHING ELSE THAT DRAWS THE CATHOLICS.... NOTHING HUMAN COULD DRAW, BUT THE MASS IS THE GOD-GIVEN SACRIFICE OFFERED THE CREATOR, IT IS HOLY THURSDAY COME DOWN AND CALVARY MADE PRESENT TODAY. MASS IS GOD REALLY AND TRULY PRESENT ON OUR CATHOLIC ALTARS, A LIVING UNBLOODY VICTIM OFFERED AGAIN

---

[391] John Allen Jr., "Coronavirus Crisis, Cardinal Pell's Acquittal, and Pope Francis's Study Commision on Women Deacons," The John Allen Show, Podcast/YouTube Video, 34:01, April 15, 2020, https://www.youtube.com/watch?v=_uEohVWP_yA&list=PL-qxMAluRCZmnSMWad-KDA5_DaSdTBIbsL&index=6

FOR THE SINS OF MEN, OFFERED, TOO, IN THANKSGIVING FOR ALL THE WONDROUS GRACES THAT UNCEASING FLOW FROM GOD'S GREAT MERCY THRONE ON HIGH.

YES, THE MASS IS THE CENTER OF CATHOLIC WORSHIP. IT IS THE MASS THAT MATTERS. WHERE THE MASS IS, THERE IS GOD HIMSELF, REALLY, TRULY, THOUGH UNDER SACRAMENTAL VEILS..."[392]

Fr. Coyle's point is that unlike any other type of religious gathering, what takes place on the altar during the Mass is a mystery beyond our comprehension—God himself is made truly and substantially present in the Eucharist at each and every Mass. Therefore, the Church holds that the celebration of the Eucharist is the "source and summit of the Christian life",[393] which is why not being able to participate in it has far more serious ramifications than what might appear on the surface. The goal of the Christian life—what being made in the "image and likeness of God" really boils down to—is that we are made for communion with God, to partake in his own divine life, the very life of the Blessed Trinity. We are made by love, for love. Heaven, oftentimes described as the "beatific vision", could also be described as simply eternity spent in the presence of God; whereas hell would be the exact opposite, eternal separation from God, being deprived of the beatific vision. The Mass—most especially reception of the Eucharist—is in one way a foretaste of heaven, an opportunity to partake of the divine life here and now. If the time we spend here on earth is meant to prepare us for eternity in the presence of God, there is nothing like the presence of God himself in the Eucharist to help do that. Even though this hardly does justice to all the factors involved, it does partially explain why the Catholic Church takes it so seriously. During the full suspension of public gatherings, Catholics were encouraged to participate in the Mass via virtual means, with countless churches and parishes

[392] Greg Garrison, "Priest supported closing churches during 1918 flu pandemic, lamented loss of Mass", AL.com, April 17 2020, https://www.al.com/coronavirus/2020/04/priest-supported-closing-churches-during-1918-flu-pandemic.html

[393] Vatican II, Lumen Gentium no. 11

live-streaming their celebrations through social media and other online platforms, but it is still not anywhere near the same.

That all being said, most government and public health officials are primarily concerned with the physical effects of a pandemic such as COVID-19, and rightly so. There is also growing recognition of the effect it is having on people's mental and emotional wellbeing as well. Unlike when the Black Death struck during the peak of Christendom, however, our modern pluralistic, relativistic society tends to relegate the spiritual to the private realm and therefore all but ignore it—something the Church simply cannot do. Striving to maintain the balance between caring for each human life in its entirety, and the danger of getting carried away focusing on one aspect alone, be it any of the above, has proved to be one of the biggest challenges the Church has had to face during pandemics up and down the centuries, the current one being no exception. As we've already seen, the way Christians responded to epidemics in the first couple of centuries quite possibly contributed to the gradual conversion of the Roman Empire and changed the course of Western civilization. But the situation has changed over the years. In a recent book, published during the midst of the current crisis, Dr. Bullivant describes the situation as follows:

> "IN THE DEVELOPED WORLD, AT LEAST, THE STATE OF OUR PANDEMIC HOTSPOTS IS NOT REMOTELY AKIN TO THIRD-CENTURY CARTHAGE OR ALEXANDRIA. THE DISEASED AND DYING ARE NOT LEFT TO DIE IN THE STREETS. ON THE CONTRARY, MEDICAL STAFF AND OTHER ESSENTIAL WORKERS ARE WORKING TIRELESSLY TO SAVE LIVES. CHARITIES, CORPORATIONS, AND SMALL BUSINESSES ARE PLAYING THEIR PARTS AS BEST THEY CAN—RALLYING RESOURCES, REJIGGING LOGISTICS OPERATIONS, AND RETOOLING ASSEMBLY LINES. THE FULL MIGHT AND MACHINERY OF THE STATE IS BEING PRESSED INTO ACTION. AS I WRITE, US NAVY HOSPITAL

SHIPS ARE DOCKED IN LOS ANGELES AND NEW YORK, AND FOOTBALL STADIUMS AND EXHIBITION CENTERS ACROSS BRITAIN ARE BEING CONVERTED INTO FIELD HOSPITALS. THE FACT THAT INFORMAL, UNTRAINED BANDS OF ORDINARY CHRISTIANS ARE NO LONGER OUR BEST HOPES OF SURVIVING THE PRESENT PESTILENCE IS ITSELF THE BEST POSSIBLE PROOF OF THEIR HISTORIC VICTORY. FOR IT IS THANKS TO THEM THAT THE REVOLUTIONARY CHRISTIAN IDEALS OF CHARITY AND MERCY—HOWEVER IMPERFECTLY REALIZED IN THIS OR THAT TIME—GRADUALLY WON OUT OVER THE PREVAILING, AND FAR MORE CALLOUS, NORMS OF THE GRECO-ROMAN WORLD. IT IS PERHAPS DIFFICULT FOR THOSE IN OUR (EVER MORE) POST-CHRISTIAN WORLD TO QUITE GRASP HOW BRUTAL LIFE COULD BE IN THE PRE-CHRISTIAN ONE. YET SO MUCH OF WHAT IS NOW TAKEN FOR GRANTED— FROM PUBLIC HOSPITALS AND HOSPICES TO FAMINE RELIEF CHARITIES AND SOCIAL SECURITY—WERE AVOWEDLY CHRISTIAN INNOVATIONS. AS BART EHRMAN, A SCHOLAR OF EARLY CHRISTIANITY (WHO IS AN AGNOSTIC), PUTS IT: BY CONQUERING THE ROMAN WORLD, AND THEN THE ENTIRE WEST, CHRISTIANITY . . . CHANGED THE WAY PEOPLE LOOK AT THE WORLD AND CHOOSE TO LIVE IN IT. MODERN SENSITIVITIES, VALUES, AND ETHICS HAVE ALL BEEN RADICALLY AFFECTED BY THE CHRISTIAN TRADITION. . . . WITHOUT THE CONQUEST OF CHRISTIANITY . . . BILLIONS OF PEOPLE MAY NEVER HAVE EMBRACED THE IDEA THAT SOCIETY SHOULD SERVE THE MARGINALIZED OR BE CONCERNED WITH THE WELL-BEING OF THE NEEDY, VALUES THAT MOST OF US IN THE WEST HAVE SIMPLY ASSUMED ARE "HUMAN" VALUES."[394]

It must not be forgotten that if these "human" values stem from the Christain vocation to love—which in turn stems from the very divine life of the Blessed Trinity—then the Church truly does have something to offer to society as a whole, and it is not limited to

[394] Bullivant, Catholicism in the time of Coronavirus, 10-12.

just its contributions to social justice, as important as those are.

In the twenty-fifth chapter of Matthew's Gospel, Jesus tells a parable about separating the sheep from the goats, assigning them to his right or left depending on their deeds. He tells the sheep, "Come... inherit the kingdom prepared for you from the foundation of the world; for I was hungry and you gave me food, I was thirsty and you gave me something to drink, I was naked and you gave me clothing, I was sick and you took care of me, I was in prison and you visited me." Asked when it was that they did such things, Jesus answers them, "Truly I tell you, just as you did it to one of the least of these... you did it to me." (Mt. 25:34-40) On September 10th, 1946, on a train ride from Calcutta to the Himalayan foothills for a retreat, the woman who became known throughout the world as Mother Teresa received what she referred to as her "call within a call"—changing her life forever. Taking those five words—"you did it to me"—literally, she began serving the poorest of the poor in the slums of Calcutta with the same tenderness as if she was serving Jesus himself. Arguably one of the most concrete examples of the Christian vocation to love put into practice in the last century, Mother Teresa was even awarded the Nobel Peace Prize in 1979 for her "work undertaken in the struggle to overcome poverty and distress in the world, which also constitute a threat to peace..."[395] Mother Teresa's "call within a call" was, in one sense, an answer to Jesus' call to metanoia—"to go beyond the mind that you have"—and embrace a new way of seeing the world. For her that meant leaving the relative safety and comfort of her convent and literally picking people up out of the gutter, exemplifying what it means to live the greatest commandment: "'You shall love the Lord your God with all your heart, and with all your soul, and with all your mind'...And a second is like it:' You shall love your neighbour as yourself.'" (Mt 22:37-39) People would often come to her asking for advice, wanting to know what they

[395] "Nobel Committee: The Nobel Peace Prize 1979 press release"

could do to imitate her. She would respond with something along the lines of: "We think sometimes that poverty is only being hungry, naked and homeless. The poverty of being unwanted, unloved and uncared for is the greatest poverty. We must start in our own homes to remedy this kind of poverty." Her point being, real Christian charity begins at home. Learn to love your neighbour as yourself.

In conclusion, even though there have been a variety of ways Christians have responded to pandemics up and down the centuries, it comes back to the same fundamental point. The Christian call is to love, to will the good of the other and actively seek what is best for them. As we've seen, this stems from the very nature of the Trinity itself, and is by no means limited to any one of the four aspects of a human person, (mental, physical, emotional, or spiritual), but is meant to incorporate all of them. The same call Jesus issued 2000 years ago in the hills of Galilee, the call to repentance, to metanoia, has not changed. Go beyond the mind that you have, and learn to see the world with the eyes of love. The Kingdom of God is at hand. In the earliest days of Christianity this meant caring for those sick and dying from a variety of diseases and illness, right up until the time of the Black Death and subsequent outbreaks of the plague. In the wake of the Spanish Flu modern healthcare as we know it began to emerge, taking its cue from Christian institutions that had already been around for hundreds of years, and in many places, have almost ended up replacing the Christian ones. The call has not changed. There are plenty of "unwanted, unloved and uncared for" people suffering from the effects of the current pandemic, whether or not they have contracted COVID-19. Start with those closest to you, and history will tell what sort of a legacy this pandemic will leave to future generations. It is being decided one act of love at a time. The Christian vocation to love changed the world once, and it has the power to do it again.

XVI.
CONCLUSION

Small-scale personally connected tragedy tends to elicit considerably more visceral emotional reactions from people than large-scale personally detached tragedy. Generally speaking, people don't react to the news of thousands of people dying in a different country in the same way that they respond to the death of a friend or loved one. To a degree, a detached emotional response to mass suffering is necessary, since reacting to every undue death in the world as if it were someone you knew would be too much emotional weight for anybody to carry. Being too detached runs the risk of losing empathy for other people and becoming desensitized to human suffering, a problem which internet news and social media feeds are exacerbating through constant bombardments of bad news and the juxtaposition of mass suffering with banal social media posts and regular news updates. Ultimately, the question of how to emotionally and personally reconcile with human suffering on a wide scale, like the question of how to spiritually reconcile, has no single correct answer. Any level of response, however, should ultimately stem from a love and compassion for other people, which in Christianity is expressed through a vocation to love. Loving other people not selfishly because of how similar they are to us but because of how different and unique they are is a huge first step to responding to any kind of human hardship, whether it directly impacts us or not.

In this book, we've looked at many examples of how people throughout history have succeeded or failed in responding to global pandemics and widespread famine. Sadly, many negative human tendencies can be seen across our discussions. The tendency for people to blame the hardship and suffering they see in their society on certain groups can be seen as far back as the 14th century in the villainizing and slaughter of Jews in response to the Black Death. Such ignorance can be seen more recently with attitudes towards Africans and gay people brought on by HIV/AIDS and in sinophobia with COVID-19. Comparisons can be drawn between the failure of some countries to respond to the COVID-19 pandemic and the refusal of certain governments to act against preventable famines. The idea that people are ignorant because God provided us with the freedom to be ignorant can be used to reconcile with why individuals act bigoted and cruelly towards other people. This idea falls somewhat short when considering evil acts such as the Holodomor, where disproportionate amounts of human suffering result from the actions of a few individuals. The ultimate question of why God would allow such immense evil and sorrow, whether it is man-made or not, has no single definitive answer. At the end of the day, the best we can do in the face of suffering and hardship is to follow the examples of the numerous good samaritans and charitable souls we have discussed, and attempt to love and help our fellow humans as best as we can.

    The Black Death, Spanish Flu, and COVID-19 pandemic highlight aspects of the evolution of the Christian Church over time. In the Black Death, we see contemporary scientific explanations fail to rationalize the plague, with religious explanations picking up the slack. This reliance on the Christian faith is reflected in the fact that religious institutions and members of the clergy were still the go-to source for healing and aid. As time moves on and scientific knowledge improves, religious explanations are not as important for rationalizing pandemics and other disasters as they once were. Simultaneously, health care systems become increasingly secularized and detached from the Christian Church. This is not an inherently negative or positive thing to happen to the Church, but simply an aspect of how Christianity naturally adapts and evolves over time. With past pandemics, we have the benefit of time and hindsight when analyzing exactly how the Church was affected. In the present moment, however, it's nigh impossible to guess exactly what the Church will look like in ten, thirty, or a hundred years, even if we weren't living in the middle of the pandemic. However, if recent history is any indicator, the Christian community will continue to be a powerful provider of charity and spiritual healing across the world.

Comparing the level of medical technology and quarantine protocols available during past pandemics such as the Black Death or Spanish Flu to the level available now, it's very easy to feel as though future pandemics will never reach the same level of mortality again. This isn't an unfair assessment to make, given just how far medicine has progressed even within the past few decades, and that the COVID-19 pandemic is so relatively small in scale and mortality compared to past pandemics. However, it is important to know that the next pandemics could be far, far worse than COVID-19. With global warming the environment will gradually become more and more suitable for diseases to live in, and the clearing of previously untouched forests to make for human development will expose people to a slew of new animal-borne viruses.[396] Air travel means that outbreaks that previously would have been confined to isolated areas and communities have a chance to easily spread to large population centres. Recently popular far-right governments such as those in the U.S. and Brazil have been shown to be inept in dealing with the COVID-19 crisis. If these governments dropped the ball this hard for a relatively "mild" pandemic, imagine how poorly they could handle a more dangerous pandemic. Looking at all of these factors, it's not hard to despair at the prospect of a second, worse pandemic. Despair, however, is unproduc-

[396] Laurie Garrett, "The World Knows an Apocalyptic Pandemic Is Coming," Foreign Policy, September 20, 2019, https://foreignpolicy.com/2019/09/20/the-world-knows-an-apocalyptic-pandemic-is-coming/.

tive. There is much cause to be hopeful. Medical technology and quarantine procedures have improved significantly over the past century, and continue to improve every day. The current pandemic has highlighted the amazing work that millions of healthcare workers, caretakers, researchers, delivery people, and volunteers do around the world. If nothing else, COVID-19 will be a teaching moment for governments and citizens worldwide. Although ideally, no "teaching moment" should come at the cost of hundreds of thousands of human lives. It is normal and natural to despair, as so many others have, in such trying times. It is in the face of great despair, however, that perhaps the most important acts of love are performed, and it is through these acts that the process of healing from suffering can begin.

www.ingramcontent.com/pod-product-compliance
Lightning Source LLC
Chambersburg PA
CBHW030117170426
43198CB00009B/646